Heterodox Economics 2

Hasan Gürak

Heterodox Economics 2

**Alternative Analysis to the Mainstream
Blackboard Economics Based on the
Concept of *Creative Mental Labor***

PL ACADEMIC RESEARCH

Bibliographic Information published by the Deutsche Nationalbibliothek
The Deutsche Nationalbibliothek lists this publication in the Deutsche Nationalbibliografie; detailed bibliographic data is available in the internet at http://dnb.d-nb.de.

Cover Illustration:
© Olaf Gloeckler, Atelier Platen, Friedberg

Library of Congress Cataloging-in-Publication Data

Gürak, Hasan.
    [Iktisat 2. English]
    Heterodox economics 2 : alternative analysis to the mainstream blackboard economics based on the concept of creative mental labor / Hasan Gürak. — 1 Edition.
        pages cm
    Translated from Turkish.
    ISBN 978-3-631-64475-1
    1. Economics. 2. Technological innovations—Economic aspects. 3. Economic development. I. Title.
    HB171.G98713 2013
    330—dc23

                                                            2013017260

ISBN 978-3-631-64475-1
© Peter Lang GmbH
Internationaler Verlag der Wissenschaften
Frankfurt am Main 2013
All rights reserved.
PL Academic Research is an Imprint of Peter Lang GmbH.

Peter Lang – Frankfurt am Main · Bern · Bruxelles · New York ·
Oxford · Warszawa · Wien

www.peterlang.de

PREFACE

This book, **Heterodox Economics-2,** consists of eight articles, which are 'complementary' to the subjects presented in the book titled **Heterodox Economics** published in April 2012 by Peter Lang International, Germany and each section, though independent, is related to the others in presenting an unorthodox and alternative approach to mainstream doctrines in economics. The distinguishing feature of all sections is the concept of the laborer's **"Creative Mental Labor"** which, more or less, has the same significance as the Marxist concept of "labor-power". The reason for this is the consideration that the laborer's **"creative mental labor"** is the source of all technological innovation which is the source of "long-term" economic growth.

Two other important and related concepts frequently referred to are **"technology-producing labor"** and **"technology-using labor"** for which we will use the shorter term **"laborer".**

These articles present alternative views but do not make any claim to have found or to reveal the **"absolute truth"** in economics. There are, in fact, no ultimate or absolute truths in any science. Therefore, some criticism of the views presented is unavoidable, in fact, it is necessary for any further ideological development.

Obviously, there will be some criticism of the views presented by the supporters of mainstream economics. Hopefully the reader, regardless of her/his ideological commitment will read and evaluate the book with an open-mind and with the same degree of tolerance and indulgence that they would afford to the orthodox or mainstream theories. I would welcome and appreciate any constructive criticism in regard to any errors or shortcomings.

I'm grateful to Prof. Dr. Cihan Dura for his priceless comments and advice on the Turkish version of this book.

And special thanks goes to John Lee from Turgutreis for his patient and valuable "mental and physical labor" in helping with the English translation.

Subject content of the articles:

The "Introduction" presents an evaluation of the present economic system, its concepts and theories.

The first article entitled **"Production Factors-Productive Factors & Income Distribution"** discusses the basic concepts and issues such as labor-laborer, the interest rate and rent.

The second article entitled **"On Value and Price"** published in 2004 in YK-Economic Review, discusses the core terms in economics.

The third article entitled **"Creative Intelligence and Technology"** consists of two separate articles also published in YK- Economic Review: The first is entitled **"On Productivity Growth"**, (1999) and the second **"Economic Growth and Productive Knowledge"** (2000).

The fourth and fifth articles are reprint translations of Chapter-5 and 7 of my book entitled **"Ekonomik Büyüme ve Küresel Ekonomi"** (Economic Growth and Global Economy), published in 2006. Both are slightly modified to adjust to the concepts outlined in this book.

The sixth article entitled "Neoclassical Marxists?" questions whether the well-known economists Lucas and Romer were "latent Marxists" due to the similarities in their approach to the concept of "human capital".

And the seventh and final article remarks on some of the negative impacts of the present economic system and offers some new insights.

Dr. Hasan Gürak

www.hasmendi.net

CONTENTS

INTRODUCTION

The richest person in this world is the one who is healthy and has true friends; not the person with highest amount of accumulated possessions.

Economic Science and the Human Being

One of the frequently asked questions of those who are interested in economic issues concerns the subject matter of economics. In other words, what is economics about? What does it investigate? And which are the main areas of interest?

But, before going any further, it would be appropriate to evaluate whether economics is a branch of the natural sciences? If it is, is it governed by the same rules that apply to other sciences such as physics or astronomy? In other words, is it possible to make observations or to measure the outcomes reliably? Can the tests be repeated under the same conditions and get the same results? And is it possible to construct "universally valid" laws?

Let us begin by introducing definitions of "science" and "technology"?

Science

Science is, according to the Merriam Webster Dictionary; "knowledge or a system of knowledge covering general truths or the operation of general laws especially as obtained and tested through scientific method". Science is systemized knowledge concerned, normally, with the physical environment and its phenomena, which can be tested using scientific methodology.

Running tests and measuring results is essential for scientific research. However, although there are some similarities, the methodology used in the physical sciences differs vastly from the methodology applied to economic research. In spite of the claims of mainstream economists, it would be a fruitless task to attempt to repeat a test "under the same conditions" as in done in the physical sciences in order to draw universally applicable laws. Because, economic events never repeat themselves exactly, except in the minds and models of the neoclassical economic academicians. In other words, there are no exact replicas of economic events. At best one can observe similarities in some trends. There is no social science including economics which is subject to pure cause and effect relationships; the concepts of "intuition", "perception" and "interpretation" play an important role in the study and in the interpretation of results in social sciences. Therefore, when studying social events, one should always be aware of the distinctive differences in the research methodology between the physical sciences and the social sciences. Perhaps, it would be more appropriate to refer to economics as the "the study of social knowledge" instead of referring to it as a "social science". However for the purposes of this study we shall continue to use the commonly accepted term "social science" throughout.

The Accumulation of Knowledge

The total amount of the already accumulated knowledge, whether it concerns the "natural sciences" or "social sciences", constitutes a "reservoir of knowledge" readily available channeled for the benefit of mankind. This reservoir of accumulated knowledge is the result of the research, discovery and inventions over a period of thousands of years. Each and every contribution was a small contribution at the time. None descended as manna from heaven as a free gift to mankind. This accumulated knowledge was created by the unique mental ability of humans. Each contribution to this reservoir of knowledge has helped to increase our understanding and control of our environment and also helped to create new useful products for the benefit of mankind.

This drive to increase our "scientific" knowledge with reference to nature can be explained under four headings:
1. to describe;
2. to understand;
3. to explain a phenomenon; and
4. to predict.

Technology

Since our concern is with economics, the discussion of "technology" seems to be more interesting than science per se. However, it is often difficult to draw a distinct line between technology and science. Technology, in general, may be defined as "the knowledge required to control and change our environment". In a narrower sense which is directly related to economics, it can be defined as "the knowledge required for the production of products"; that is "**the knowledge of production**" or "**productive knowledge**". It can be regarded as the application of knowledge for commercial purposes in contrast to scientific knowledge which has no direct relationship to commercial interests.

The sole purpose of an enterprise making use of the available technology is to maximize its long term profits. In other words, the main difference between scientific knowledge and productive knowledge is that productive knowledge is "profit driven". In modern societies, technologies are generally invented, to maximize profits. In a dynamic process, new technologies are constantly being introduced employing the successive stages of research, invention and innovation which lead to the introduction of either "**new processes**" or "**new products**".

Sometimes a new technology in the form of a "new process" reduces the unit production cost which may or may not, lead to a price reduction. The most striking examples for this kind of new technology can be seen in the computer or mobile phone sectors. Sometimes a new technology introduces entirely new products which are often accompanied by a new production process. Such technological developments exert a positive impact in the markets which keep up the demand high for. As a result, the pessimistic predictions of some economists on the decline of the profit rate and economic stagnation in the long term never take place. In other words, the major determinant factor of this lack of decline in profits and the absence of stagnation is the continuous introduction of new technologies which are derived from human "**creative mental labor**".

Creative Mental Labor & Technological Progress

The single most important input of production which leads to long term welfare growth is productive knowledge i.e., technological progress. As we emphasized previously, new technologies are the products of human creative mental ability. However, the creation of a new technology alone is not sufficient to secure welfare growth. There is also a need for an appropriately qualified labor force in order to make efficient use of the new technologies. Access to a qualified labor force is as important as the creation of a new technology because in the absence of an appropriately qualified labor force new technologies cannot be used efficiently.

Software

The "software" sector has been one of the fastest growing sectors in the last two to three decades and this process seems likely to continue. Software can be described as the indispensable intermediary tool to make computers and electronic devices functional and operational. In other words, software is an intermediary input required to make a computer or an electronic device work. Each computer or electronic device requires an "appropriate" software configuration to become functional and to meet the needs of user.

Software can be analyzed in two groups:

1. Software specific for a particular device ("**device-specific**" **software**):
2. Software for the communication between the device and the user ("**user-specific**" **software**).

The first, i.e., "device-specific" software is designed to make a device functional. It prepares the device for the user by setting into motion various internal operations and by facilitating links between them. It displays the icons on the screen and prepares the system for further instructions from the user. A computer or an electronic device may sometimes require more than one software package in order to boot up and prepare the device for the individual use.

The latter, the "user-specific" software can be further divided into two major groups:

2-a: **General "user specific" software**, e.g., Windows, Games, Powerpoint, etc..

2-b: **Commercially "user specific" software** which is software designed specifically for a bank or a trading firm, or an assembly line or for the exchange markets, etc.

Figure-1 shows us the annual receipts and payments of the 'US General Use Software' in years 2007 and 2008. According to the data supplied by the Bureau of Economic Affairs, the receipts of the USA increased from about $ 29 billion to about $ 32 billion which shows a $ 3 billion increase. The payments abroad also increased by about $ 3 billion, increasing from about $ 26 billion to $ 29 billion. This data gives us a clear picture of the growing importance in global trade of software transactions.

As the BEA data shows clearly, both the value and the volume of the global trade in software has been steadily increasing. The expansion of the "device-specific" software sector is closely and directly associated with the expansion of the computer and electronic device sectors. However, the analysis of this important and constantly growing sector will not be dealt with in this study. The reason for this is that "device-specific software" for the computers and electronic devices are considered to be "complimentary" components. To put it another way, this type of software is normally introduced into market along with the hard-ware, i.e., the goods. Therefore, it seems to be sensible not to consider the device-specific software as a product independent from the main product.

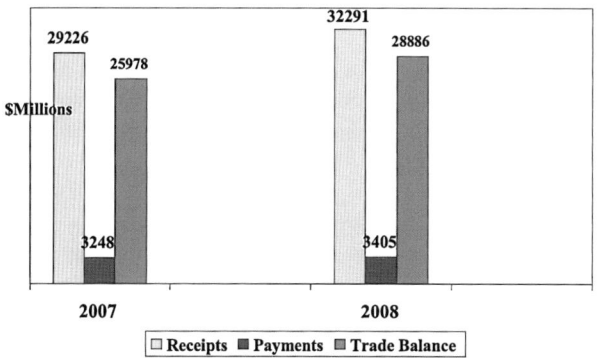

*Total for affiliated and unaffiliated

U.S. Department of Commerce | International Trade Administration

Figure: 1 Source:
http://web.ita.doc.gov/ITI/itiHome.nsf/ea087b1279d5fb1985256ce00053c33c/
8f32a883f9024ddc85257359004b2775/$FILE/Software%20Trade%20Data%20Tutorial.ppt#
256,1,Software Trade Data Tutorial, 2012-05-30

On the other hand, the 'user-specific' software displays some differences from the 'device-specific' software. For example, 'user-specific' software prepared for a 'specific' enterprise can be purchased on the market as an 'independent" component of the hardware. In other words, the 'user-specific' software can only be used by the "specific" user which now gives it a monopoly control over this specific software. Therefore, the "user-specific" software is not subject to trade in the market like any other product which is normally subject to competition.

Although we are aware of the importance of both the "device-specific" and "user-specific" software trade in the global market, we shall not deal with either of these issues at this particular juncture.

A Brief Evaluation of Economic Ideologies

The beginning of economic relations can be traced back to a period when man began to produce products in excess of his basic needs. With this surplus of products they could enter into an exchange relationship with others who also had some surplus which was in demand by other consumers. Economics started to become accepted as a separate discipline in the late 18th century when Adam Smith's works came to the fore about 230 years ago. We shall follow suit and accept Adam Smith as the founder of modern economics with the introduction of his well-known book entitled *"An Inquiry into the Nature and Causes of the Wealth of Nations"*. As the title of the book says, it was about "The Wealth of Nations", i.e., how the wealth is created and distributed amongst classes, as opposed to todays neoclassical approach which analyzes "distribution" in terms of "scarcity".

Adam Smith's economic system was first challenged by David Ricardo who made significant contributions to economics especially on the subjects of value, functional income distribution and international trade. However, Adam Smith's as well as David Ricardo's ideas were blown apart shaken from their foundations by Karl Marx in the 19th century. The neoclassical theories which came into being in the 1870s came to the rescue of the then economic theory which had been under attack from Marx. Since then, some economists began to believe that economics is a "science" just like astronomy and physics with "universally" applicable laws. The neoclassical school came a long way in refining and developing their doctrines and models which are largely based on the Newtonian laws of physics. Nowadays, it is accepted that Newtonian physics as the benchmark for scientific method has been developed and changed substantially since Newton's time. However, neoclassical economists still seem to be loyal to the Newtonian scientific methodology.

One of the basic principles a student of economics learns from the mainstream textbooks today is that markets are functioning "perfectly"; there are countless small producers who have no power to influence the quantity supplied or to determine the price, technology used is the same and available to all. The present "scientific" system alludes to no historical events which might have influenced today's conditions. Everybody behaves rationally making decisions based on "perfect" knowledge. The consumers are "selfish", "non-emotional" and "rational" in the sense that they always attempt to maximize their own benefits. Human psychology plays no role in the economic life and behavior of the robotic "homo economicus". All these features belong to the "imaginary world of neoclassical doctrine" and are a fallacy.

However, it would be a mistake to dismiss the neoclassical doctrine entirely on the ground of it being a fallacy. The doctrine can make useful contributions to economic science if it is treated as a "normative theory" instead of a "positive theory" as the neoclassical economists claim.

Since it is not our purpose to evaluate neoclassical theory, let us continue our investigation into the "subject matter of economics".

The Subject Matter of Economics

According to the mainstream economic introductory textbooks the subject matter of economics is defined as **"the allocation of scarce resources"**. That is to say, it claims that man's "needs" are unlimited while the resources required to meet these needs are scarce. Thus it is the task of economists to secure and allocate these scarce resources in an efficient way. Let us consider this proposition closely to see whether it is true.

First of all, it is not "the needs of man" but the indoctrinated "wants" of an insatiable man, inflated and promoted by the "wants of commercial enterprises", that is unlimited. The basic needs of man can generally be described as, the need for nourishment, the need to find shelter to live and the need to protect his/her life. Mankind cannot survive if these basic needs are not met. There is, without doubt, a close relationship between the economic-technological development level of a community and the quality as well as the quantity of its basic needs. For example, it would be irrational to expect people to live in primitive dwellings when they have the opportunity to live in much better quality residences equipped with modern household appliances. Similarly, the quality of food, drink and clothing are naturally much higher today than in the past. Eating delicious food with a knife and fork and living in a well-designed and decorated house are now considered among the basic needs depending on the development level of community, though they may be described as basic needs of a "secondary grade".

What about the physical consumer items such as TV sets, mobile phones, computers, automobiles and service items such as touristic trips, listening to a musical, or watching a film? Can they be considered as "the basic needs" of man? The answer is certainly, NO! These are goods and services which make life easier, more comfortable and pleasant but cannot be classified as "basic needs". They might be beneficial in many respects but life can go on smoothly even without them. The demand for such consumer products depends on the prices of products and the purchasing power of consumers which are in many cases influenced by commercial advertising. For example, there is a continuously growing demand for the services of "new generation" mobile phones and tablet computers which are marketed as a "natural part of modern life" so if we don't have one, our quality of life would be lower. Nowadays it certainly appears to be a "natural" demand to have access to contemporary communication items.

However, "not" possessing these items would not make life miserable or intolerable. The demand for such consumption items are generally "created" and "inflated" by commercial advertising and engenders a "more and more" attitude towards consumption within the system.

Let us return to the subject matter of economics, whether it is the allocation of scarce resources, as the neoclassical textbooks claim. The related question that pops immediately to mind is: **What are these "scarce resources"?** If it is the **"number of laborers"** that is scarce, the remedy is at hand: eliminate all barriers to worker mobility, the problem is solved. If the reference is to the scarcity of **"qualified laborers"**, encourage policies aimed at improving the skills of the labor force; the problem will be overcome sooner or later. Let us assume that the problem is the scarcity of **"capital-goods"**. No big problem. By increasing the supply of capital-goods, the problem will soon wither away. It would be highly irrational to assume that **"capital in the form of money"** is scarce, because there is more than enough money circulating in the global financial markets to make (not to earn) more money from money. Besides, there is always the option to increase the money supply by printing money or by using the existing banking methods.

If none of the "supplies" mentioned above is scarce, then the concept of "scarcity" might be referring to the availability of natural resources. If so, then we should be looking for the appropriate methods to allocate these scarce natural resources in an efficient and fair way to all countries. Although the scarcity of natural resources is a serious global economic problem, it does not seem to be what the neoclassical textbooks refer to. Well, **just what are these "scarce resources"** that constantly trouble the minds of mainstream economists?

There does indeed seem to be a scarce production resource which should not be scarce at all; "technology". Technology was broadly defined above as; "the knowledge required in order to control and to change our environment", and in a narrower sense, as the "productive knowledge" of production. As we know, productive knowledge e.g., technology, is produced by the creative mental abilities of the human mind. As a product of the human mind knowledge is accumulated over thousands of years by millions of marginal contributions to the "reservoir" of knowledge. As such it should be available to all mankind without restriction or at least with only a limited level of restriction.

Could it be that **technology**, i.e., **productive knowledge**, owned like a "property" or a "commodity", the use of which is restricted by patent rights, is the **"scarce production resource"** for which we are searching?

If technology is this critical scarce resource, then we may find the roots of many major economic problems in imperfect technology markets. Accordingly, the solution can be found by doing away with these imperfections in the technology market.

Actually, the invention and development process of a new technology is a costly process, especially in dynamic sectors such as bio-chemistry, avionics, computers and genetics. However, once it is produced, the distribution costs of a new technology in the form of "knowledge" are quite low, if not zero. Yet, the system used in dealing with patent rights of new knowledge makes the availability of it to other producers impossible unless the "owner of technology" decides to share it. The right of patent entitles the owner of the new technology to exclusive privileges in the market until the other enterprises catch up, if they can. Meanwhile, the owner can exert a monopolistic power, determining both the price as well as the quantity supplied in line with his or her own interests.

Yet, as we know, no knowledge drops from heaven and no new "productive knowledge", e.g., new technology is produced from the "exclusive knowledge" of the technology-producer. Any new knowledge that is created is always a marginal contribution to the reservoir of knowledge accumulated over thousands of years. In the absence of such an immense reservoir of knowledge, no one would be able to make the new inventions we frequently come across every day. Therefore, granting exclusive privileges to an enterprise to make "monopoly profits" and to control the markets seems unfair and needs to be revaluated on "ethical grounds".

And that's not all there is to it. Any new knowledge created and patented is due partly to contributions from the society in which it is developed. Because, if the society were unable to provide the appropriate technological infrastructure, the educational facilities and the general economic and financial environment, the development of many new technologies would not even have been imagined in the first place. All these factors involved give a society "ethical" right to make claims on the new technology and the privileges granted to it.

In short, given the facts stated above, economists should start reconsidering the exclusive patent right system on ethical grounds in regard to all those who contribute to a new technology. If exclusive privileges are going to be granted enabling monopolistic profits, then all the parties concerned should get their "just and fair" share.

The Subject Matter of Economics - Reconsidered

Economics is a social science studying all kinds of economic relationships within the society. To be more specific, economics is about every economic issue ranging from the supply of products, to consumption, employment, inflation, trade, etc. with the aim of solving problems, maintaining stability and predicting future developments. A broader scientific definition should cover all kinds of micro and macro-economic transactions, any cause-effect relationships in economic events, welfare growth and the elimination of any actual or potential problems. The purpose of all these efforts is to maximize the economic interests of human beings and to secure a prosperous environment. Therefore, human beings are always at the center of any economic system; everything is for them.

A glance at actual economic transactions clearly shows us that contemporary man often behaves quite "irrationally" in terms of his sequence of priorities, where, often, material interests take precedence over human values and ecological balance. There is no doubt that the economic system in which we live has a lot of influence over the formation and the implementation of "values". However, one should not forget that we created this system; it has not fallen as manna from heaven.

Whose Interests Have Priority in the System?

As mentioned before, the main objective of the economic system is to maximize the income and welfare of men but of which men exactly? More specifically, to which economic group's or class's interests should priority be given?

Should it be to;
1. The capital owners who bring into being production? or
2. The working people, who constitute the majority of the population?

Alternatively, should society give priority to the capitalists whose only interest is the maximization of profit, or to maximize the interests of those people who actually add value to the products that are supplied, i.e., the laborers?

Historically, we observe that the political sector has always been fed by the economic sector. In other words, the political and economic sectors are always in cahoots with each other. The more economic power one economic group gains the more political power it can get, control and direct. The capitalist's rising economic power in the 17th and 18th century had gradually taken over the Mercantilist's political power by reshaping the political, legal and bureaucratic structure in line with their economic interests. Since then, the wheels of the economic order have been spinning in order to serve primarily the interests of the capital owners.

In the past, laborers gained some improvement in their economic conditions in lieu of their increasing political power. However, the actual economic and political power of the capital owners is still much greater compared to the working people's economic and political power which in turn reinforces the dominance of the interests of one relatively small segment of the population. For ex-

10

ample, in case of an economic downswing, the group of people who suffer most is the laborers. There is no doubt that capital owners would be affected during a period of economic turbulence but not to the same extent as the laborers would be affected. Obviously there are always exceptions to the rule. The natural outcome of the economic order in which we live, the system is one of competitive survival which eliminates the weakest and allows the strongest survive. However, the number of working people negatively affected by an economic downturn is and always has been greater than the number of the capital owners.

Mandeville and the "Fable of the Bees"

The Fable of the Bees tells that if each individual behaved in a selfish, self-interested and immoral way, the benefits to society would be maximized.

The Fable of the Bees was first published in 1705 and aimed to reveal the then prevailing conditions in England using the metaphor of a beehive. The striking feature of this beehive was it provided the premises for plenty of immoral behavior. The most striking and frequently observed immoral behaviors were forgery, arrogance and a passion for luxury goods. In fact, all the trading sectors were involved in some kind of immorality.

The political system of the beehive was like monarchy in England. Lawyers were cheating their clients. Doctors were charging a fee from the patients they couldn't heal. Similarly, merchants were cheating their customers, judges accepting bribes and the Ministers of State were embezzling from the State Treasury. The population of the beehive was, without exception, arrogant and had a large appetite for the luxurious life.

The reader expects that all these immoral behaviors would, in time, be revealed and the bee population would repent, suffer the consequences and eventually find the "right" path: But, this expectation remains unfulfilled. On the contrary, all these kinds of immoral behavior bring happiness and welfare to the beehive population as a whole.

Mandeville points out that there is a choice to be made between moral virtues and economic development; both cannot profitably work together.

Adam Smith had first severely criticized the Fable of the Bees in his book "Theory of Moral Values" but later he adapted a similar line of thinking to Mandeville on the grounds that human beings are, by nature, selfish and self-interested as stated in his book: "An Inquiry into the Nature and Causes of The Wealth of Nations".

E.B. Ruben, 2011, İktisadın Unuttuğu İnsan, p. 23-25; Translated by H. Gürak

---

When the economy moves upwards, profits rise first before real wages follow suit later, subject to the negotiation power of their unions. If the union's wage claim is above the acceptable limit by capital-owner, there is always a risk that jobs may be lost due to what is called "the runaway process" of the enterprises. Production continues in countries where wages are lower, supported by generous financial incentives which help to increase profits. Briefly, the present economic order favors one economic group over the other.

In passing, there is an important issue that should not be overlooked in regard to the performance of the enterprises; i.e., technological progress. It is the innovative oriented enterprises which survive the competition in the long term, cet. par. Those who technologically lag behind are normally doomed to a gradual process of extinction. The key word of survival is "technological progress".

In our age, new technologies are usually developed, if not invented, by the R&D departments of the major enterprises, especially in regard to the high cost technologies. As mentioned elsewhere, all technological progress is the result of "creative mental labor" but the benefits accrue, to significant extent, to the capital owner who buys and thus owns the patent rights of the new product. In a time of economic straits, even the creator of the technology may become superfluous to the interests of the "owner" of the technology. This is how the system works.

Economic Power & Behavior

Consumers are an essential and indispensable part of the economic system leading to its stability and sustainability. After all, they are the customers, i.e., purchasers of the commodities and services supplied by enterprises in their pursuit of profit. If they decide not to purchase a product or postpone the purchase decision, the system becomes unstable from the ground up.

The critical question that springs to mind in relation to the potential power of the consumer is; to what extent are they aware of their potential power and its impact on the economic system? Or, similarly, to what extent are they able to use their potential power to maximize their own interests?

As we were informed by the international media toward the end of 2011, a group of people initiated a movement with the well-known slogan "Occupy the Wall Street" in New York City. They claimed that they represented 99 percent of the population. However, a short while later they found out that, contrary to their expectations, the actual support for their demands and political power they thought they had behind them wasn't as large as they expected it to be. There was neither 99 percent support for them nor for the changes did they hope to make. The important lesson to be drawn from this action is that in economics the "potential" power of the majority is not always equal to the "actual" power held by the minority.

Some people in some countries make brave attempts to form a "new kind of economic behavior" with the slogan; "do not purchase more than you need". The idea behind this is to make life simpler by curbing the desire for unnecessary consumption. No doubt this kind of behavior would not please the producers and sellers of the various consumption products. It wouldn't be a surprise, if the enterprises in some countries denounced this consumer behavior as "perverse" or "hostile". Because such an action could lead to a decline in consumption which in its turn would lead to a reduced quantities supplied and to a decline in profits. For instance, if an individual shares a car with others to drive to work or to go shopping this behavior would sooner or later reduce the demand for cars, fuel, insurance, etc., which in turn would reduce the total amount of profit. By participating in these "perverse" behaviors individuals could salve one's qualms of conscience but definitely not the conscience of the producers and sellers of the products involved.

Being satisfied with "enough" goods and services and sharing one's possessions that are in "excess" of one's personal needs with others seems like a delightful and pleasurable and self-satisfying human behavior. But, the related critical questions are: How much is "enough"? And, how much is "in excess" of one's personal needs? Another related and highly critical and emotive question is: How conscious are people that their behavior is self-indulgent and is used to maximize their own personal feelings of self-gratification?

Each individual instinctively thinks that one's behavior is both "rational and conscious" and would naturally refuse to accept that it was otherwise. Yet, as we know from experience, human beings do not always make conscious and/or rational economic decisions in their regular life. In fact, irrational economic decision-making and the resulting behavior is not at all uncommon in economic transactions. For instance, we often observe that the people who consider themselves to be conscious and rational consume unhealthy drinks from a can; as we know cans pollute the environment. Another example of this irrational and unconscious behavior is the well intentioned sharing of unhealthy processed foodstuffs with those suffering from hunger which often causes more damage to their health in the long term and has little impact on alleviating their hunger in the short term. It is a prime example of "the road to hell being paved with good intentions". Or, not objecting to tin production on the grounds that the sector creates employment would lead to an undermining of the sustainability of the ecological system and as such cannot be considered as either a "rational" or a "conscious" economic behavior.

Two important questions arise, in conjunction with "conscious behavior", which require elaboration. The first one is: What is "consciousness"? How do we define "conscious behavior?

According to the definition of the Institution of Turkish Language "to be conscious" refers to the **"ability of men to apprehend himself and his environment"**. Based on this, we can define the "conscious economic man" as the **"man who does not only act rationally but is also aware of the consequences of his/her economic actions"**. He/she anxiously avoids being the cause of any kind of intentional damage to other individuals as well as to the environment. Certainly anyone can make unintentional mistakes. After all, we are not perfect. But such unintentional mistakes will be considered as exceptions and will be overlooked.

To summarize, conscious men are those who are aware of their responsibilities and of the consequences of their economic transactions while paying due respect to the environment. Our "conscious man" is substantially different than the mechanical "homo-economicus" who is "highly ambitious, economically rational and competitive" as proposed in the neoclassical doctrine. Our conscious and rational man is aware that money is an important ingredient of the economic system but spends his/her income and wealth "consciously".

Being Selfish vs. Unselfish (Benevolent)

One of the distinguishing features of man is his selfishness which means that man is concerned excessively and exclusively with himself/herself and is intent on maximizing his/her own advantages, or his/her well-being. As the matter of fact, that when evaluating a person with respect to the commonly accepted "moral values", it is a greater virtue to be "benevolent" or "unselfish". But, the concept of "moral value" is a subjective concept. Something that is "moral" for a person or group in one community may be diametrically opposed to those of another community. There is no way to measure the value of a moral act nor does one expect a profit in return for being "moral". In other words, benevolent persons do not expect to gain any personal advantage from their benevolent behavior. Therefore, their behavior is the opposite of the expected "rational economic behavior" of man as described in neoclassical doctrines.

The best reward "the benevolent" gain seems to be the emotional appreciation of others. This "benevolent" behavior or the "good feeling" that it engenders is an uncommon feeling and a non-existent behavior for the "neoclassical homo economicus". For the efficient functioning of the neoclassical economic system the desired and appropriate individual behavior has been defined as selfish and economically rational. For the neoclassical system to be efficient, it has to provide for the maximization of profits and personal gain based on "economically rational" and "competitive" decisions. Selfish individuals who strive to maximize personal advantage are most welcome in their system, while the unselfish and benevolent ones are "undesirable", to say the least. For example, not too long ago large families consisting of parents, grandparents and children were not uncommon. But, such large family structures are not welcome for the profit maximizing economic system because they share many things in the house ranging from furniture, kitchen ware, TV-sets to food, cars, etc. This is not desirable for the profit maximizing motive because a smaller family means a greater consumption of all items; furniture, car, TV-sets, computers, refrigerators, dishwashers, carpets, etc. Therefore, small size families are preferred and encouraged by a system whose virtues are indoctrinated by so called "social organizers" or "community engineers". There are countless examples of this type of behavior in the developed countries.

Now, let us take a closer look at the expected hypothetical behavior of a family living according to neoclassical criteria. By assumption, the family consists of a mother, father and a child. The family always makes "economically rational" decisions, and plans to buy a car worth 50,000 TL. But, to their surprise, they find out that their child suffers from a sickness and has to undergo an operation immediately which costs exactly 50,000 TL; the exact amount required to buy the car they so passionately desire. If the operation is postponed the child's life will be jeopardized. What should the "neoclassical" family decide? Should the family spend their savings on the operation? Or should they decide to buy the much desired car?

Everyone who thinks like a normal individual would expect the same humanitarian and moral decision: "Definitely the money will be spent on child's operation". But would this humanitarian and moral decision be an "economically rational" decision for our neoclassical "homo-economicus"? After all, there is the risk that they may not be able to save the same amount of money required to buy the car. Yet, having another child virtually costs nothing.

The selfless actions of animal-lovers' are another example because their benevolent and unselfish behavior contradicts the expectations of neoclassical values and interests. Because these "irrational" individuals according to neoclassical values spend their money and time on something that does not bring any material gain. If they spent their money on goods and services, their behavior would be classified as "economically rational" a lá neoclassical criterion.

Fortunately, the neoclassical style of selfish and economically rational behavior has as yet not been able to best this benevolent behavior either permanently or decisively.

Concluding Remarks

An important issue discussed in the previous sections was the subject of economics and whether economics should be considered as a science like the natural sciences. The conclusion was that economics is a "social science" studying human economic relationships and transactions. As observed, "human beings" appear at the center of all economic analyses and every economic decision primarily aimed at improving the economic well-being of man.

The importance of the role of technology or to be more specific "technological progress" was emphasized as a basis for all long term economic relationships and their ensuing economic transactions. It was also pointed out that all technological progress which continuously improves material standards is the product of human "creative mental labor". In other words, all new products that make life more comfortable and easier are created by "creative human labor". There are peaks and troughs in the direction of welfare growth but it is a fact that so far this long run process has always trended upwards. Therefore, the pessimistic predictions of some economists have never become true. Neither has the long term growth rate halted nor have long term profits fallen towards zero. Nor have we ever observed the imaginary "state of equilibrium" over any sustained period in the economy. There have of course been coincidental "selected and frozen" periods in a constantly dynamic economy.

The single most important input for welfare is "knowledge" or more specifically "knowledge about products and processes". However, the existence of a knowledgeable labor force, i.e., a qualified labor force, is as important as the knowledge input in order to secure the improvement in the state of material well-being. In other words, in the absence of "appropriately qualified laborers" a society would be unable to produce "new technologies" and/or unable to make use of them efficiently. Therefore, the single most important input in production relationships appears to be the "qualified labor force", i.e., the source of all knowledge and the user of it.

In short, the human being is at the center of any economic analyses, all the time, everywhere and at each stage of development. But the economic system as such seems to give priority to the economic interests of only a relatively small group of human beings; the capital-owners. The economic interests of the largest group, the workers, seem to be secondary in regard to economic decisions and the advantages gained which is a consequence of unequal distribution of economic and political power.

Based on the "selfish and economically rational" decisions of the individual, the main expectation of the producers and sellers is that all the products supplied should be consumed which would maximize their incomes and profits. The more the consumers demand, the higher the profits will be. Therefore, a greater and an unlimited demand is encouraged and promoted through various channels which trigger the consumers desire for further consumption. Slogans such as; "needs are unlimited" or "this product will change your life" are all used to inflame the consumption fever of "selfish and economically rational" individuals. From the point of producers and sellers these methods have so far been successful. But, such trends do not seem to have made people happier.

It is high time to reassess our values, behaviors and priorities and pay more attention to "high moral values".

References

Fromm, E.	2003	Sahip Olmak ya da Olmak
		Arıtan Yayınevi, İstanbul. Çev. Aydın Arıtan
Gürak, H.	2006	*Ekonomik Büyüme ve Küresel Ekonomi,*
		Ekin Kitabevi, Bursa.
--- " ---	2011	*İktisat,* Genesis Kitap
		Ankara.
Ruben, E.B.	2011	İktisadın Unuttuğu İnsan
		Bağlam Yayıncılık, İstanbul.
Smith, A.	1976	An Inquiry into The Nature and Causes of the Wealth of Nations, Vol. 1 & 2

1- PRODUCTION FACTORS, PRODUCTIVE FORCES & INCOME DISTRIBUTION

> The composition of this book has been for the author a long struggle of escape... –a struggle of escape from habitual modes of thought and expression. The ideas, which are here expressed so laboriously are extremely simple and should be obvious. The difficulty lies, not in the new ideas, but in escaping from the old ones...
> J.M. Keynes, 1973, General Theory... Preface

Introduction

Every introductory textbook on economics refers to some key and fundamental concepts like "factors of production (**FoP**), and income distribution", among others. The orthodox theories inform us that there are two factors of production; **labor** (**L**), and **capital** (**K**), but often fail to specify what they really are or what they really imply. But, once upon a time, in the earlier texts of Classical economics, there used to be a third factor of production, **nature** (**N**) widely known as land. Some contemporary textbooks claim that there is even a fourth factor, the **entrepreneur** (**E**). In the post-1950 era, the number of the factors of production appears to have increased. There is nowadays another factor influencing production "**re-discovered**" by Solow; namely **technology** (**Te**). In addition, since 1960s, there is yet another factor which has an effect on production; "**human capital**" (**H**).

The definitions of concepts like capital, labor, human capital, productive factors, and production factors are dubious and, require revaluation and re-examination. Accordingly, widely used analyses of economic relations are also dubious and, require reassessment and a more candid explanation.

The widely used and accepted definitions found in textbooks and among scholars are mostly of the Neoclassical heritage, which appear to represent "**a playground for academic economics**" rather than a realistic approach leading to a better understanding of "actual" economic facts and relationships. Therefore, in regard to the present global economic (dis-)order, there seems to be a need for a new as well as a more realistic approach from alternative angles and if and when necessary "new" definitions of "old" concepts.

The Purpose and the Subject

The purpose of this work is partly to reconsider and when necessary, to redefine, "conventional" key concepts such as labor, capital, interest, rent, and accordingly analyze the distribution of functional income. Thus, the questions to be analyzed in the following sections are:

- What are the factors of production?
- Is the term **production factor** identical with the term **productive factor**? What is the difference between them?
- Which factors "**earn**" a share of the income generated? And which factors receive an "**unearned**" share?
- Which forces influence the distribution and redistribution of income among labor and capital; the ability to generate value or the ownership rights?

Related to the above questions addressed in the discussion are:

What are wages? Why are wages paid? And what is capital? Is it simply a certain quantity of "hoarded" money or does the term refer to the "capital goods" (**KG**) of production, or both? Why does "capital" always strive to get "profit"? Is capital a "productive" factor in production like nature or the services of the laborer? What indeed is "Human Capital"? Is it labor or capital? Which factor makes the greatest contribution to the growth of productivity[1] and thus to the generation of income? What is the nature of incomes such as interest and rent, which accrue from money holdings?

A distinctive feature of the analysis will be the distinction between "**productive**" factors and "**production**" factors. This distinction will enable us to better grasp **"the relationship between production and the distribution of functional income"**. Such a distinction is necessary in order to understand the contributions and ensuing rewards of the factors employed in the production process. To be able to say whether the income is distributed "rationally" and "justly", we have to know the specific contribution of each factor to the added value of the product. In this function the "added value" is denoted by the term "VA" (= π + **LWC**).

The central hypothesis in this paper is that "mental labor assisted by physical labor"[2] is the source of **all value-created or added** "given" nature as the indispensable and irreplaceable supplier of "**all initial use-value**"[3]. Nature which

---

1   For the definitions of terms like "productivity", "productivity growth", "productivity increase", "technology" see Gürak 2006.

2   Physical labor is the simple eye-hand coordination of laborer requiring no mental labor.

3   For a discussion/evaluation of the terms see Gürak, 2000-a; 2000-b; 2004-a; 2004-b.

supplies the raw use-values[4] (utilities) which are transformed[5] into exchange-values[6] by the efforts of the "mental and physical labor services"[7].

In fact, this article is built on the previous work found in previous articles (see Gürak-1999, Gürak-2000-a and Gürak-2004-a).

The basic assumptions, unless otherwise stated, are:
- No inflation.
- No government intervention.
- Prices reflect the market values of products[8].

Factors of Production (Fop) & Productive Factors (PF)

In general, there is no distinction in the standard mainstream introductory economics textbooks between the two similar yet "factually" distinct terms; "factors of production" and "productive" factors. The concept we are accustomed to is that of "factors of production", as both labor (L) and capital (K) are assumed to be "productive". Following the new developments after 1950s, new factors of production such as "technological innovation" (A) and human capital (H) began to appear in both textbooks and economic models. The distinction between the two concepts presented here are, however, somewhat different. Though the concepts might "appear" to have the same implications, they are fundamentally different. As will be seen below, a proper distinction is not only necessary but also a prerequisite for a sound economic analysis.

Factors of Production (FoP)

According to the conventional economic textbooks, there are only two **FoP** these are **K** and **L**, used in the production function (**Q**) as defined below:

$$Q = f(K, L)$$

The standard textbooks, unfortunately, provide us with no globally acknowledged definition of capital. They hardly ever specify what it implies and almost never question the nature of capital as to what it actually is or how it has

---

4   Use-value: utility enjoyed by the end-user from a product.
5   Transformation implies the change in original formation/position by some labor effort.
6   Exchange-value: value created by labor for trade (exchange); or the relative worth (price).
7   Labor service is the mental/physical capabilities of laborer in the generation of value.
8   Product: tangible commodities as well as intangible services.

emerged in the first place. "Capital is a factor of production" and that is the end of it. Apparently, it is a "sacred" concept that is beyond any doubt and must be accepted without examination or analysis as "a priori". At least for the neoclassical doctrine and its proponents.

The definition of labor, too, is far from satisfactory, though less cumbersome than capital. In economic models, labor sometimes refers to "uneducated and unskilled" laborers and sometimes to laborers with average skills and education. Whenever the concept of "human capital" is used, it refers to the former, i.e., unqualified laborers without any education. Yet, the concept of labor also refers to the work done by laborers[9].

In reality, there are more inputs of production than the two suggested above. **Production factors** are all the prerequisite inputs of production to produce anything with an exchange-value. In analogy, they are like the ingredients of a soup. Without the necessary inputs (ingredients) you cannot make the soup. The **FoP** (ingredients) comprises inputs ranging from raw materials to energy, semi-finished inputs, plant site, etc., as well as labor services. For instance, thousands of inputs of varying qualities are required for the production of airplanes or computers. On the other side, a teacher or a consultant in the service sector seems to perform their tasks with much less physical inputs. Accordingly, a house painter or a shoeshine boy requires considerably less inputs of production than the producers of computers or cars.

The most frequently required inputs of production are:
- Mental and physical labor services.
- A production site (factory, workshop, etc.).
- Raw materials.
- Energy.
- Semi-finished inputs.
- Management services.
- Marketing and distribution services.

We can categorize the inputs of production into three groups:
1. Capital (K).
2. Labor-laborer (L).
3. All other inputs ranging from raw materials, semi-finished inputs, energy, etc.

Regarding the aim of this work, there are two critical and vital questions that need to be answered:

---

9   We will discuss the concept labor in detail in coming parts.

1. Can a **factor of production be** a "**productive**" factor at the same time, capable of generating or adding value?
2. Is capital a productive factor[10]?

Productive Factors (PFs)

The answer to the first question above regarding the factor of production being "productive", the neoclassical doctrine would claim, without hesitation, that capital is a productive factor. They would argue that the marginal productivity analysis of capital displays this feature clearly.

It is true that some factors of production display a distinct feature by being "**productive**", i.e., capable of generating or adding value even in the absence of other inputs. But capital is definitely not one of them. Nor are other inputs such as energy, tools, intermediary inputs, with the exception of labor services, "productive" in any sense.

There are only two kinds of "productive" forces fitting the definition:
1. Nature[11] (**N**), i.e., the supplier of all use-values (utilities) in the form of the raw (unprocessed) inputs of production; and
2. Labor effort (**L**), i.e., is the "only" factor capable of adding value to the original use-values supplied by nature.

The fundamental difference between the **FoP** and **PF** is the ability of the latter to produce value. Every **PF** is, at the same time, a **FoP**, but every **FoP** is not a **PF**.

Nature: A "Factor of Production" as well as a "Productive" Factor

It is a commonly acknowledged fact that nature is capable of providing the inputs, i.e., the raw materials, for any production even in the absence of any other assistance, including any intervention from the laborers. This feature makes nature, beyond any doubt, a "**productive**", factor. There are thousands of kinds of life arising and surviving due to the productivity of nature. All kinds of species on earth, even human beings, the most advanced species of all creatures, take their nourishment from the gifts (the products) of nature. There would be no form of life without the productivity of nature. Thus, that nature is a "produc-

---

10   This is one of the well-known "Cambridge controversies".
11   Also, and frequently, referred to as "land".

tive" factor is beyond any doubt or discussion in the sense that it provides the physical inputs with use-values (utilities) for the benefit of mankind.

The end-users, in our case human beings, consume the products of nature, not only to survive but also to produce various processed products. As long as the products of nature are used for personal or family consumption, they contain "use-value" only for the end-users. For instance, apples on a tree indicate **use-value (utility)** if used for personal consumption. But, if they are brought to the market by an effort of the laborer such as their collection and transportation, they contain an "exchange value".

For example, say that Deniz, the father of family which lived in ancient times, collects some fruits, the "free gifts of nature", to feed his family. The fruits contain "use-value" (utility) for him and his family. Assume that Deniz possesses more apples than he requires and decides to exchange the surplus apples for oranges. The apples they consumed had a "use-value" for him and his family but the surplus apples which he exchanged for oranges had an "exchange-value". These are the concepts and relationships often cited in the works of the Classical economists, which show us that "exchange-value" arises only when some labor-effort is expended on the product. Briefly, nature provides things with use-values, which are either consumed directly by end-users or used as exchange-values involving the use of some "labor time" spent on the product.

In addition to being a productive factor, nature also provides all kinds of inputs required for many different outputs in the form of raw materials. Therefore, nature is not only a productive factor, but her products are the indispensable inputs of production in the form of raw materials. In the absence of the raw materials as the input factors of production, there would be no physical items reshaped by labor for consumption, such as computers, cars, houses, TV-sets, etc.

Ecological Balance and the Sustainability of Production

The ever increasing numbers of advanced commodities offered to the end-users have their origins in nature in the form of raw materials. But, unfortunately, the generosity of the nature is not unlimited and unless the natural resources are used with caution, care and consideration, they can be completely exhausted or become unusable over time due to relentless usage and wastage. A polluted and eventually dying nature would mean an end of civilization and technological advancement, regardless of any developmental level.

Labor (Laborer): A "Factor of Production" as well as a "Productive" Factor

Let us recall that in orthodox economic books, there were only two factors of production.

$$Q = f(K, L)$$

where **L** denoted labor, e.g. "i" the quantity of laborers[12] (i=1,2,...,n). The textbooks normally do not specify the features of **L**; whether it is trained or educated, if so, in what capacity, the degree of talents and experience. But whenever **H** (human capital) is introduced into the analysis, the ensuing and logical conclusion would be labor is untrained and uneducated, and endowed solely with basic eye-hand coordination. That means **L** does not possess any inherent abilities or experience, save simple eye-hand coordination.

Regardless of the level of the abilities of the laborers, labor is normally regarded, by the orthodox neoclassical doctrine, as a "**productive**" factor of production receiving a share of income in accordance with its marginal productivity. The role played by this marginal productivity according to neoclassical doctrine is open to discussion. But, the fact that labor is productive, i.e., capable of producing value, is a fact beyond any doubt. Labor services can "**generate or add value**" to the supplies of nature by transforming (reshaping) them, either for personal use or for trade. Among all the inputs of production, only the labor services are capable of "adding value", given nature's supplies (see Gürak, 2000-a; 2004-a; 2006).

At this stage, it would be useful to discuss some related concepts, such as labor (mental and physical), labor-services, laborer, and labor –force, before proceeding further with the analysis.

Some Related Definitions

Given the human capital **(H)**, according to some economic textbooks and models, **L** refers to "i" the number of laborers endowed solely with eye-hand coordination. Although many analysts are well aware of the fact that **H** and **L** are actually two sides of the same coin and that no labor, especially in our time, is without some degree of formal or informal training and/or education, they continue to treat **H** and **L** as if they are two distinct factors of production. This "ideological" ignorance is not only confusing but also leads to an incomplete and unsatisfactory economic analysis.

---

12   Laborer is used as synonymous to worker or employee.

Yet, the word "labor", literally, implies **the execution of work by a laborer**. It is the result of the mental and/or physical efforts of a laborer using given "inputs". In other words, labor is the exertion of mental and/or physical capabilities of the labor to execute a task, i.e., to produce "use- and/or exchange-"values. In short, labor is the result of work by a laborer.

Laborer (L) is, in fact, the able-bodied individual, who possesses some **mental as well as physical abilities**. He/she is the person who carries out the work or task by using his/her mental and/or physical faculties to generate use- and/or exchange- values. To put it differently, the laborer is the human being who is hired by the enterprise in exchange for a wage, i.e., he/she sell their services in the market place and **"earn"** wages in return. Wages are not paid for hiring the laborer for his/her personality or body as such, but for the "services" to be obtained in order to realize the production.

Mental labor (L^m) refers to the faculties (abilities and skills) of the human mind developed through formal and/or informal training and/or education as well as the experience gained, given the persons' natural skills. Creative mental labor is the source of all present and future technology. As we have discussed elsewhere (Gürak, 2000-a; 2000-b; 2004-a; 2006), given nature's supplies, mental labor-services are **the genesis and a continuous source** of all products and thus of economic growth and prosperity.

Physical labor (L^p) is the simple eye and hand coordination of the able-bodied person. The execution of physical labor requires, by definition, neither special training, education nor any mental abilities. In reality, there is no such thing as "purely" physical labor; it exists only in theory. Every labor possesses some degree of mental ability. Even choosing the right electric button between, say red or white, requires some degree of mental ability.

The term **Labor force (LF)** implies the total quantity of the laborers (able-bodied persons) willing to sell their labor services in return for wages. Economic models normally use **L** in the same sense as **LF** without making a distinction.

The Concept Labor (Laborer) (L) - Reconsidered

A proper definition of labor (**L**) is necessary for a sound analysis. In this work, **L** refers to "Laborers" willing to work at the going market wage rate. **L** does not only comprise able-bodied persons with eye-hand coordination, for there is no such thing in reality. Each **L** embodies, in addition to physical abilities, some degree of mental ability which is acquired and developed through formal and/or informal training and/or an education enriched by experience. As stated before, the term labor refers to the "laborer" or "worker" or "employee".

L is "not" a commodity like a TV-set or radio bought and/or sold in the markets. It is **the only source of any added value** and **the ultimate consumer** of the value produced, along with other income obtaining groups, of course. And these features make **L** a rather unique factor in production. Commodities belong to their owners, but **L** is only hired for labor services for a given period of time in return of wages. The ultimate control of labor services belongs to the labor. Otherwise, it would be slavery.

Since the labor is endowed with different skills, abilities and experiences, the wages paid by end-users in return for labor are not, and cannot be the same for all. In other words, the qualities of labor are heterogeneous, and so are the wages. As the terms "quality", "faculty", "education", "ability", "skill" and "experience" are all "subjective" concepts. A sound measurement and eventually any inter-firm or inter-country comparisons of labor services are cumbersome, if not downright impossible.

Is it possible to measure the contributions of the laborer accurately? According to the proponents of neoclassical doctrines who consider neoclassical economics as a science with "universally" applicable laws as in physics and astronomy, everything can be measured with certainty given some "basic assumptions". They would not hesitate to make attempts to measure a subjective analytical concept such as "utility" in economic models, though they had given up the measurement of cardinal utility long-time ago.

Throughout this book, subjective concepts like quality, faculty, education, ability, skill, and experience will not be attempted to measure for measurement of labor services is considered rather cumbersome, if not impossible.

To summarize; taking into account related discussions elsewhere (Gürak, 2000-a; 2000-b; 2004-a; 2006);

1. **L** is one of the indispensable and irreplaceable factors in the "factors of production" (**FoP**);and
2. **L** is also an indispensable and irreplaceable "productive factor" (**PF**).

Capital: Is it a "Productive" Factor?

The first and major problem with the concept of "capital" arises with its definition. There seems to be no "globally" agreed definition of "**capital**". For some, it refers to "**capital goods**" implying machinery and perhaps tools. For others, it refers to all the inputs of production, ranging from buildings and energy to semi-finished inputs, except for labor services. For some, capital is the total value of assets used in the production process, including labor services.

Since the concept of capital is one of the key concepts of economic theories, it would be useful to take a look at it.

Historical Perspective

The concept of "capital" has been one of the most frequently used and discussed concepts of all economic theories. According to Hausman who made a comprehensive study of the subject: **"Economists possess no good theory of capital and interest."** (Hausman, 1981, Ch.10) The theories possess elegant models and theorems but they, he said; **"... do not enable one to explain real phenomena of capital and interest"** (Hausman, 1981, Ch.10) as they fail to grasp the essence of the issue.

For some "Classical" economists the concept of "capital" covers not only the "capital-goods" but also the laborer's wages as well as all the other inputs of production. Thus, according to Ricardo "capital" is;

> "... that part of the wealth of a country which is employed in production, and consists of food, clothing, tools, raw materials, machinery, etc. necessary to give effect to labour." (Ricardo, 1990: 95)

For Marx whose views resemble Ricardo's, capital is the sum of money which contains some specific features.

> "... in itself this sum of money may only be DEFINED as capital if it is employed, spent, with the aim of INCREASING it, if it is spent expressly in order to increase it". (Marx, 1976:976)

Accordingly, the owner of the sum of money is called a "Capitalist". However, according to Marx' definition, the petty-bourgeois or owners of small-scale businesses would not qualify for the title "Capitalist" although the petty-bourgeois had the same goal as the true capitalist; that is increasing and/or maximizing the initial sum of money employed to produce commodities, i.e., making profits.

The well-known economist of the Classical period J.S. Mill had a similar definition of capital like his contemporaries.

> "What capital does for production is to afford the shelter, protection, tools and materials which the work requires, and to feed and otherwise maintain the laborers during the process... Whatever things are destined for this use... are Capital." (in Schumpeter, 1954: 634)

So called mainstream economists seem to prefer a different approach and assume that capital, as one of the main inputs of production, is a "scarce" input. According to this approach capital became another "productive" factor along with labor and thus was entitled to receive "a reward" called "profit". The competing Austrian School of economics objected to this "scarcity" approach claiming that capital cannot be a "productive" factor, but mainstream economics overcame the challenge and it is nowadays common to encounter a "scarce factor" approach to capital in contemporary economic textbooks.

For Marshall, capital was an outcome of the productive employment of the laborer and "waiting". Accordingly, capital should not be considered as a product of the labor only for it would lead **"... by inexorable logic to admit that there is no justification for interest"** (Marshall, 1961: 687). Marshall had to introduce a new definition of capital which said: **"... a store of things, the result of human efforts and sacrifices, devoted mainly to securing benefits in the future rather than in the present"** (Marshall, 1961: p.787). Thus, the capital-goods produced by laborer would be entitled to some profit, if there were not consumed immediately but employed in the production of future benefits, i.e., as a reward for "**sacrifices**".

In modern theories, it is customary to treat "capital" as a **productive factor** just like labor, which entitles the owner of the capital to an income; i.e., profits. This approach regards "capital" as being capable of producing or adding value to a product, which is greater than the initial investment. Thus, as a "productive" factor, it **earns** (receives) income in accordance with its marginal productivity.

Is Capital a "Productive" Factor Capable of Generating Value?

In order to provide a reply to this critical question, three alternative definitions of capital will be reconsidered below.

1- Capital á la Classical School: Classical economists had defined capital as the financial funds used to initiate production, i.e., to bring together the **FoP,** with the purpose of producing commodities with exchange-values. This definition does cover "**all**" the inputs of production ranging from raw materials to plant site, semi-finished inputs, machinery, tools, energy, as well as the labor services. Capital can also be defined as:

$$K = FC + VC + LWC$$

FC denotes all prearranged and invariable inputs of production like plant site, machinery, etc., **VC** denotes all the tangible and intangible inputs ranging from raw materials, to semi-finished goods, energy, etc., varying along with the quantity of output, and **LWC** denotes the total wage cost of employees (**LWC=w*L**) [13].

Capital á la Classical School brings together the necessary inputs production, including the services of the employee, with the specific purpose of making a profit. By doing this, the capitalist takes the risk of losing the whole or some of the capital outlaid. But all the inputs can only transmit value, which is equal to the sum of all values of the inputs consumed. For instance, assume FC= 40 TL, VC= 30 TL and LWC= 20 TL, the value the output contains is 90 TL. There is no additional value produced by the capital, meaning that the capital defined á la Classical School is **NOT PRODUCTIVE**.

2- Capital goods (KG): KG, or synonymously, the physical implements of production ranging from tools to advanced machinery are, in fact, nature's supplies transformed by labor services to increase the "value-adding ability of the labor(er). Since the value of the **KGs** is given (known) in advance, they produce or add no value at all but only transmit the value already embodied in them. Their price embraces values transmitted during their production process plus any profits. The sole purpose of the employment of **KGs** is to increase the productivity of the laborer. Thus, **KGs** deserve compensation only to the extent of the value transmitted to final product. For example, if the value of a capital good is 100 TL and totally exhausted after one use, the contribution is worth just 100 TL. No more, no less. Anything exceeding the compensation for transmitted value is neither economically rational nor justifiable. Therefore, **KGs** are a **FoP** but not **PF**. In other words, **KGs** are **NOT PRODUCTIVE**.

The "capital-intensity" or rather the "technology-intensity" of production is another controversial concept, which refers to the degree of the exploitation of the technology embodied in the "capital-goods". The more advanced the technology, the more qualified the employee is likely to be in order to realize (to effect) production. In contrast, less advanced technologies are likely to require a relatively higher number of employees with relatively less qualification. Therefore, it would be more appropriate to distinguish between these production methods as "technology-intensive" and "employee-intensive" instead of "capital-intensive" versus "labor-intensive" as is at present customary. The degree of technological progress does not, in any way, alter the position of **KG**; they still cannot transmit any more value to or cause depreciation of the final product than the value they initially possess. In other words, the variations in the degree of technological advancement embodied in the **KG** do not make them more or less productive.

---

13   w = wage of employee and L=the number of employees. All incomes-expenditures are gross values, i.e., no taxes, unless otherwise stated.

3- Hoarded money or monetary capital: A clear distinction exists between the "hoarded money" and "invested money" (Capital á la Classical School). A person can possess large sums of monetary funds, e.g., savings, but that does not make him an entrepreneur (a producer of goods or services) but just a money-holder. Such savings do not produce any value although they might grow by the amount of interest. Savings can be associated with value-generation only as "invested money" (Capital á la Classical School), that is if and when employed in production of exchange-values by combining the employee services and other inputs of production. In other words, although invested money (Capital á la Classical School) is an essential **FoP**, it is **NOT** a **PF**. In other words, the money capital is **NOT PRODUCTIVE**.

To sum up, the foregoing analysis indicates that, in spite of the claims of Neoclassical doctrine and its marginal productivity analysis, capital is not, and cannot be a **productive factor.** Capital-good(s) is (are) employed to increase the "**productivity**" of labor services. Thus, they can, at best, be described as "**productively employed**" in the process of production.

Capital Reconsidered

For the supply of goods and services with an exchange value, the entrepreneur has to combine certain production inputs such as labor, production plant, energy, raw materials, semi-finished inputs, tools and machinery. In order to combine these inputs, the entrepreneur has to have access to **assets for investment** which is "a prerequisite" of production. Throughout this work, all the assets required to enable the output of goods and/or services shall be called "**production capital**", or simply "**capital**", unless otherwise stated. In other words, with the employment of capital the entrepreneur creates a "**productive capacity**", which is not money, but is expressible in monetary terms. Capital invested endows the labor with production facilities and inputs with the goal of increasing the value of the original amount invested, and will be denoted by the letter "**K**". In the absence of capital, a combination of physical inputs together with the services of labor could not take place. In other words, production goals could not be realized.

To put it differently, **K** represents the stored-up values of the inputs of production including labor. As J.B. Clark pointed out, it is "expressible in money, but not embodied in money". When employed, it places the **factors of production** at the service of the labor, which is **the only value-adding productive factor.** According to this definition, neither "hoarded" money (savings) nor the "capital-goods", i.e., buildings or land fit our definition of **K**, which designates "**value**", not quantities. Given technology, **K**, consists of the following:

K = FC + VC + W
FC = plant + machinery + tools
VC = raw materials + semi-finished inputs +energy + etc.
LWC = labor wage costs (wages * quantity of laborers)

Origin of the Capital-goods

As shown elsewhere (Gürak, 2000-a, 2004-a, and 2006), given the output of nature, the labor is the generation of all the value added by virtue of its mental and physical abilities. The major role is played by the former, i.e., mental labor, which contributes to the growth of wealth by improving the productivity of present labor assisted by past labor embodied in the implements of production, i.e., transformed natural products. Assume a barter-exchange economy consisting of two persons with no implements of production (no tools). And further assume that the mental labor of worker produces a new technology, which introduces a primitive tool which increases the productivity of the worker. That tool is a **capital-good**, a transformed, i.e., re-shaped, natural product. Accordingly, all the inputs of production, except for raw materials, are transformed natural products by the workers. The labor-time spent in the past in transforming the original natural outputs into physical inputs, are stored in the "**means of production**", i.e., "**capital goods and tools**", in order to assist the productivity level of the "present" workers.

Let us attempt to find out the origin of capital-goods in a simple production equation:

$$Q_t = f(K_t, L_t)$$

L, denotes labor, **K**, capital-goods, **Q** output quantity and **t**, time. Let us take any capital-good, say a drill-machine and analyze its composition. The drill-machine is made of various physical components assembled by some form of labor. So, the capital-goods K_t embodies both K_{t-1} and, L_{t-1}. Taking a closer look at K_{t-1} will show that it is not a pure physical input either, for it embodies some form of labor, as well.

$$K_t = f(K_{t-1}, L_{t-1})$$

As the backward composition analysis continues, labor services will appear at every stage of production.

$$K_{t-1} = f(K_{t-2}, L_{t-2})$$
$$K_{t-2} = f(K_{t-3}, L_{t-3})$$

At the end of backward process, there will be only two inputs left; natural input and labor-services. Adding new factors such as technological change (**T**) or human capital (**H**) to the production function would not affect the argument on capital. In short, capital is not a "productive" factor while the creative mental labor is the source of all **KGs**, given the natural products.

Is Capital a Scarce Factor of Production?

Given the amount of globally available financial resources, it would hardly be logical to argue in favor of the scarcity of capital in terms of financial assets. Financial assets can only be described as "abundant" rather than "scarce".

On the other hand, one can argue that the developing countries do not possess sufficient amounts of capital-goods and that is why capital-goods can be considered as scarce inputs in production. However, given the abundance of global financial funds, there should not be any problem in purchasing the capital-goods required for production, if funds are made available. In other words, if sufficient quantities of global financial assets are made available for production purposes wherever required, there would not be a scarcity problem regarding the acquisition of capital-goods.

Scarcity of human resources with the necessary qualifications to make the best use of capital-goods seems to be a greater problem; in fact, it is maybe the major impediment to economic growth. Imperfect global technology markets and the transactions taking place therein seem to be another major problem area with serious global implications. "Scarcity of technology", or rather global restrictions to accessing technologies, especially in developing countries, seem to be a major impediment.

Other Factors of Production

So far we have studied only three factors, nature (**N**), labor (**L**) and capital (**K**). Yet, as previously mentioned, there are many more factors (inputs) involved in production, especially in products with advanced technology such as airplanes and computers. The critical question is; are there other factors in production, which are, **productive** just like **N** and **L**, and capable of producing and/or adding value?

The answer is a very short, clear and definitive: **NO!**

There are "no" other **productive factors** other than **N** and **L** capable of producing or adding value. All "other inputs" are the products of nature transformed (re-shaped) by the efforts of labor. The inputs of production **transmit value** to the product as much as the value consumed during the process of production. If, say, 10 TL worth of steel is used, then the value transmitted to the new product is just 10 TL, nothing more and nothing less.

Human Capital: Laborer or Capital?

According to the jargon used by economists, like T. Schultz, G. Becker, R. Solow, R. Lucas, P. Romer, G. Mankiw and many others, there is, in addition to the factor "capital" in the conventional sense, another meaning of the capital concept which is vital and fundamental for production; "**the human capital**", designated by the letter **H**.

Lucas (1988), assigns **H** the role of "**the sole source of all economic growth**". For Romer (1990), it is the source of all technological innovation, thus of increasing returns and continuous long term growth. The so-called **K**-models refer to both, the "orthodox" meaning of the term "capital" as well as the "human capital".

What is human-capital? What is the difference between conventional meaning of "capital" and "human capital"?

"**Human capital**" simply refers to the mental abilities and capabilities of the laborer acquired through education and experience. It comprises the skills, abilities and "learning-by-doing" as well as the experience of the laborer. In other words, it designates the quality of the mental abilities used as inputs provided, by the laborer over and above simple eye-hand coordination. It does not make reference to the laborer himself but to the quality of abilities demonstrated by the laborer. It is: "**The skills and abilities possessed by an individual, which enable him/her to earn an income**[14]".

The factors influencing the magnitude of **H** are:
1. Official school years attended; and
2. Learning-by-doing; or
3. Both.

However, there are two additional and extremely important factors influencing the emergence of human capital; these are;
1. Experience; and
2. Informal education gained from the socio-economic environment.

H is, normally, acquired and advanced through both formal and informal education dependent on the natural skills of an individual flavored by the experiences that they acquire over time.

---

14   The Penguin Dictionary of Economics

Human Capital & the Laborer without Mental Abilities

The concept human capital refers to the "mental abilities" of the laborer, or, alternatively, to the **skills of the laborer**. The critical question is; is there any laborer without some degree of human capital? To put it differently, is there any laborer or part of any laborer, which employs eye- and hand coordination, only?

The answer to this question was given above in the section on labor. The same answer is valid; NO! There is no known labor-force or part of a labor-force that relies solely on eye-hand coordination. That is to say that every laborer possesses some degree of mental ability along with simple eye-hand coordination. Because each movement of one's muscles receive an order directly from brain, the command center of all mental activities.

If human capital is the skill or the ability of a laborer, why, then, is the term associated with capital and not with the laborer and defined as **mental labor**? Is the choice of term, "capital", instead of, for instance "laborer", merely a coincidence?

The reason seems to be ideological rather than scientific. Any association with the laborer could jeopardize the century long struggle of Marginalism to survive Marxist revolutionary ideological attacks on the capitalist system. Changing the expression from "Human capital" to "Mental labor" would produce different conclusions. As a result, the sacred world of the neoclassical temple may encounter serious questions which would lead to the shaking of its sandy and dubious foundations.

Types of Incomes[15]: "Earned vs. Unearned"

According to the analysis in previous parts, there are only "two" **PFs** capable of producing or adding value on their own and thousands of **FoP** used as mere inputs in the supply of useful goods and services. **FoP** adds to the value of products only as much as they transmit. For instance, if 10 TL worth energy is used, the amount of value transmitted by energy to the final value of product is 10 TL, no less, no more. The same applies for all **FoP** including the **KG**. Assume that the following values apply for one unit production of good **X**;

LWC = 150 TL (**L*w**)

15　Income and wealth distribution before and after production takes place has always been one of the most vividly discussed subjects of economics. Both are, in practice, unequally distributed in all market- or capitalist economies. We will assume that initial income distribution is exogenously given.

KG = 100 TL; depreciate totally after one use only.

VC = 200 TL (energy, semi-finished goods, etc.)

TC = 450 TL.

P^x = 500 TL (=**TR**)

π = 50 TL (TR – TC)

The amount of value transmitted by the inputs of production is 450 TL, but the product is sold for 500 TL. The value-added or the income generated, with the supply of product **X** is 200 TL.

$$VA = LWC + \pi = 150+50 = 200 \text{ TL}$$

How is this income to be distributed?

In our sample, the distribution is self-evident; 75 percent (150 TL) goes to the labor(er) as wages and 25 percent (50 TL) goes to the entrepreneur (the capitalist) as profit. But in reality, there are more than two income-receiving groups. Although it is not mentioned in the sample above, some share of the **VA** (income generated) accrues to other groups in economy as **interest** and **rent**.

Below, before going into income distribution analysis, an attempt will be made to describe four types of incomes and their nature. Incomes will be categorized as "**earned**", as a reward on **productive economic activities** generating value, and "**unearned**", as return on **unproductive economic activities**. The analysis will help us to see whether the income obtained is economically rational.

A Brief Reminder on "Productive" Factors

There were only two "productive" factors of production, nature (**N**) and labor (**L**). **N** is, beyond any doubt, the genesis, the spring of all commodities and the "other" productive factor. But as such, nature herself does not receive any part of the income (**VA**) generated in return of her generous supplies being used as inputs, as she is not an entity with economic interests like the other productive factor, human beings. In the sense of economic transactions, she provides "**use-values**", only. All she requires is adequate care and feedback in order to continue her sustained productivity.

1- Wages - "Earned Income"

The owners of labor services, e.g., laborers, **earn** wages for the contribution, as a "**productive**" factor of production. The verb "earn" implies that the factor makes a distinctive contribution to the generation of value as a productive factor. To put it differently, the laborer embodies certain abilities and skills enriched by training/education, which shapes the capacity to labor. As a result of setting this capacity to perform a specific task, some value is generated. In return, the laborer earns a living by renting out his/her labor-time for a price commonly referred to as wage. The value generated by labor(-er) is either used for personal use (designating use-value) or exchanged for other products (designating exchange-value).

Since the issue of how wages of labor services are affected by improvements in technology in the long run was discussed elsewhere (Gürak; 2006), it will suffice to indicate that the nominal wage rate is determined after individual or aggregate negotiations with employer(s).

2- Profit: – "Earned" or "Unearned" Income?

Profit motive is the long term driving force inducing the entrepreneurs to engage in the supply of products ranging from the basic items of life to high-tech commodities and services. Therefore, the rational objective of any enterprise is to generate the maximum possible profits. In case of declining profits, survival of the enterprise would be jeopardized. In the short term, the enterprise may pursue other temporary objectives than profit maximization such as acquiring a targeted market share, keeping the shareholders satisfied, maintaining the status quo or reaching a sale target. But, there is always only one motive in the longer term, profit maximization.

In the pursuit of the profit maximization, the enterprise makes some direct and indirect contributions to the community such as generating funds for taxation, new jobs, promoting economic growth and improving the living standards. But, all these useful contributions of an enterprise are the byproducts of profit maximization motive. An entrepreneur does not set up an enterprise with the purpose to generate employment or to pay taxes or to develop a region or community, but to make profits. Without any profit motive, the enterprise may as well slash its own throat as it would cease to exist.

In spite of its crucial importance for the system, there seems to be no consensus on the nature and origin of profit, just as it was the case in the definition of "capital". During the early stages of capitalism, the Classical economists were unable to draw a distinct line between the profits and interest. According to Schumpeter, Adam Smith might be credited with two different theories and Ricardo with three or even four; 1- abstinence, 2- residual, 3- entrepreneurship and 4- unpaid-labor. But, he said; **"it is more realistic to say that they had no definite theory at all."** (Schumpeter, 1954: p.648).

More than a century ago, James Mill (1821) and McCulloch (1825) had treated profits as **the wages of accumulated labor**. Capital goods were treated as "accumulated or hoarded labor", thus going on "earning" wages, e.g., profit. Analogously, the maturing wine in the cellar was pointed out as earning wages (profits) for the owner as the time goes on. According to this approach; "**... capital goods are the result of saving"** and; **"any net yield of these capital goods is in the nature of a payment for the service rendered by saving"** (Schumpeter, 1954: p.659). For Cassel, profit was the price of capital disposed.

For Marx, profit meant simply the surplus value, e.g., the unpaid labor. According to the unpaid labor version, the labor services were employed; say for 10 hours per day, but received wages for only 8 hours. The difference worth two hours' wages was unpaid labor time accruing to capitalists as profits. That difference also indicated the exploitation of laborers by the capitalist, according to Marx.

The abstinence theory treated profit as the return on the service rendered by saving, e.g., the price of saving. Abstaining, i.e., postponing present consumption implies some **risk** that the postponed consumption may never be realized in future. If there is indeed a postponement of present pleasure in expectation of more future pleasure, as in the case of "atomized" and "unsaturated" small investors, the argument might be justified. But, in case of the "saturated" entrepreneurs in terms of consumption, the abstinence argument would lose whole ground. It would be hardly reasonable to argue that Rockefeller, DuPont or Sabanci family members are actually postponing their present pleasure when they invest. Why should there be a profit reward if there is, in fact, no sacrifice or postponement of present consumption? After all, abstinence does not generate or transmit any value to products.

According to Schumpeter, profit was the price of potential capital while it meant, for Keynes, the price of not hoarding inducing the capitalists not to keep liquid funds, e.g., savings. For a businessman, profit is the difference between total revenue and total costs, including the payments for employees, R&D costs, capital goods, raw materials, etc.

The predominant neoclassical doctrine does not offer a satisfactory theory, either. It tells us that profit is the return of the marginal productivity of capital but there is no globally acceptable definition of capital even among its proponents.

Profit and the Risk Factor

Given the costs of production, the price paid by end-users for products normally exceeds the total value transmitted by the inputs of production. The difference between total revenue and total costs is the **"profit"** (π) paid to entrepreneur for the **"risk"** assumed. Risk can be defined as the probability of obtaining less than the value of capital employed in production. It involves the partial or total loss of the savings employed for the output of commercial products. There can be several reasons for the appearance of risk:

- Miscalculation of market response (insufficient demand).
- Unanticipated economic obstacles (customs duties, tax rates, crises).
- Unanticipated non-economic obstacles (political instability, war, etc.).
- Competitors' behavior (price-battle, new products, etc.).
- Management errors (wrong decisions).

The risk factor does not transmit any value to the product itself, i.e., it does not generate any value. But the profit as return on risk for employed savings is a necessary prerequisite for the efficient operation of the system. In other words, profit is the return on total investment capital to undertake production process, subject to risks. Thus, profit arises as a result of the economic activity to supply useful products.

Assume an economy consisting of two individuals; **X** and **Y**. In time, individual-**X** saves some of his/her weekly income, say 20 TL. Instead of consuming it immediately and after a time-span of, say five weeks, he/she would accumulate savings worth 100 TL. Assume that individual-**X** uses the savings to combine hired labor services with other inputs of production (invests) to supply a commodity. Given demand, the new investment by individual-**X** would increase the total output, thus the total wealth while, probably, generating employment for additional manpower. The individual-**X** makes a clear contribution by taking the **risk** of losing some or all of her accumulated savings and thus "**earns**" an income, i.e., profit. It would be not only economically rational and justified but also necessary for the promotion of further investments. In the absence of profit as reward, there would be no incentives to invest. In short, **profit is paid for the risks assumed**.

A Distinctive Feature of Profit

There is a rather distinctive and controversial feature of profit.

Profit is not some value generated by the entrepreneur. It is a value transfer from end-users to capital owners in excess of the total value transmitted by inputs of production. In other words, it is a transfer of purchasing power from the buyer to seller in excess of the cost of production, which is given by the predetermined values of inputs, including wages. The amount of purchasing power transferred is shaped by the supply-demand conditions and competition.

3- Interest - "Unearned" Income

Throughout history, interest has always been a frequently discussed and controversial subject of economic transactions. There was time when interest was forbidden by religion as well as by civil law, even in developed countries. Today, interest is still outlawed and considered as a great sin in countries ruled by Islamic principles. Yet, in spite of all unfavorable and not infrequently hostile attitudes, interest has been able to survive all challenges. Nowadays, billions of dollars circulate in global and domestic financial markets to make the best use of financial opportunities, to breed money by money transactions.

The fact that enormous amounts of hoarded money obtain income in the form of interest does not make it immune either from criticism or the questioning of the role of interest in the proper functioning of an economic system. Since it is an acknowledged fact that the rate of interest plays a significant role on the allocation of credits in both the production and the consumption of goods, its detailed and careful analysis is essential for the effective and productive functioning of the economic system in which we live.

As shall be seen below, the interest mechanism as it functions today to large extent functions to the detriment of the rational and efficient allocation of financial resources and against the general interests of global economies in terms of increased output and employment.

What is the Interest Paid for?

Let us begin by making a definition of interest. There have been different definitions of interest throughout history and there still seems to be no generally recognized definition. Once, Classical economists like A. Smith, Ricardo used to refer to the return on invested capital as interest. And again, some Classical economists considered interest as the price of savings; e.g. money holders by postponing their present consumption to a future date make a sacrifice and deserve interest on their savings. For Keynes, interest rate helped to determine the liquidity preference level. According to another view, interest is the price of the money loaned. And, there are also various definitions of interest by Islamic economists.

Here, we shall define interest as the "monetary return" in excess of the loaned money, e.g. income as **return on money loaned**. In other words, the income not originating directly from the output of goods or services is **"unearned income"**, that is interest. According to this definition, any monetary debt payment exceeding the original amount loaned, say 100 TL, is interest payment.

There are two distinctive features of interest that differ from profit:

40

1. There is no direct supply of goods or services.
2. The risk assumed by the lender is normally in proportion to the securities given.

The discussion of interest will be taken up from two different angles:
1. As the return on the **unproductive employment** of savings.
2. As the return on the services supplied by financial institutions (**FIs**) using the services of labor and subject to competitive risks.

Monetary savings can be used:
1. **1-To earn** income (profits) if **productively employed** (invested) to produce goods and services, which is called **capital**.
2. **To receive** income (interest) if employed on "**unproductive**" assets, such as banking deposits, bonds, obligations.
3. **To remain idle** (**unproductive**), under the pillow or in safe box.
4. **To promote output** if spent for consumption.

We have already discussed the first item capital and its revenue profit above. What we are interested in, now, is, the second, **interest** on savings not employed **productively** as capital, e.g., income on **unproductive** economic activity. Before proceeding further, a distinct line between **productively employed** savings and **unproductively employed** savings will be drawn, below.

Productive and unproductive employment of savings

In contemporary economies, savings (financial assets) not employed as capital, i.e., savings not productively employed to produce goods and/or services, often appear in the form of financial assets like bank deposits, obligations or bonds. As distinct from capital, such assets do not make any direct contribution to output or income generation, at least not directly. Interest on unproductive savings seems to arise from the mere ownership of hoarded money. Such funds, not directly involved in the supply of products, abstain from assuming the related risks of supplying products. Yet, nevertheless, they provide an income to the owner by increasing the original amount of hoarded savings.

The distinguishing characteristics of interest on unproductive savings are:
1. It is a "predetermined" rate, independent of production.
2. It has no relation to supply of products, at least not directly.
3. Money produces more money.

It is said that savings such as bank deposits, bonds, etc., do contribute to output "indirectly" by supplying funds for producers. It is true that there is such an "indirect" association with investment and output. But, nevertheless, it is not the holders of savings who assume the risks of production, but the intermediary financial institutions. The money holders make a deal with the financial intermediary only on the rate of interest, regardless of the outcome of investment. In case of bankruptcy of the producing firm, which is not an infrequent event, the money holders, are not directly affected.

Since hoarded savings are not directly related to the supply of goods or services, a line separating it from the productive employment of savings is not only proper, but also a necessity. That is why such savings are referred to as unproductively engaged savings and the return on it called **"unearned"** income. The risk the owner of unproductive savings takes is of a speculative kind common to financial markets. Risks such as high inflation reducing the real rate of interest obtained or bankruptcy of the financial institute.

Since money holdings as such are not capable of adding or generating any value, the critical question is:

Is interest on hoarded money economically rational, with regard to its contribution to increased output, employment and prosperity, or is it rather a detriment to them?

Promoting Productive Investment

Policies promoting or encouraging the **productive employment** of savings and discouraging income generation from **unproductive employment** would be to the benefit of the rational allocation of financial resources and increased output as well as employment.

Utilizing unproductive savings: A hypothetical case

Assume an economic system where all kinds of interest on **hoarded money** is forbidden by law, cet. par. What would be the likely economic implications? Would this new situation lead to a diminished amount of investments caused by the shortage of assets for investment? Would future economic progress and prosperity be at stake?

On the contrary, quite the opposite would be true.

As rational economic agents, normally, show a tendency to increase and maximize personal gain by seeking the economically most profitable areas for their savings, there would be four options for money holders, in the absence of the interest on hoarded money:

1. **Invest** personally in productive economic transactions, i.e., supplying goods and services.
2. **Invest** in already established firms by purchasing shares.
3. **Keep the savings idle**, thus receiving no income.
4. **Spend** on consumption.

In the first two cases, the money holder(s) would be subject to competitive risks, employ labor(er) and probably abstain or postpone the enjoyment of present consumption. As a result, not only the individuals but also the aggregate economy would benefit. If the third option is preferred, say as a precaution or to avoid risks, then the money holder would be not only be abstaining the enjoyment of present consumption but also assume the risk of the diminishing purchasing value of savings in case of inflation or currency devaluation. The fourth option, spending on consumption, would contribute to increased output and employment.

A rational economic behavior for a money-holder hoping to increase savings would be the preference of one of the first two options. If the hoarded money is encouraged to be placed at the disposal of productive investment, the accruing returns would be both, economically rational and qualified as **earned**. The principle task of an efficient and rational financial system ought to be channeling the savings into **productive** employment.

Income of the Financial Institutions on the Services Supplied

With regard to interest obtained by financial institutions (**FIs**) one has to look at the type of service provided before categorizing the income as **earned** or **unearned**. If the financial resources placed in **FIs** are used in promotion of supply of goods or services, then they are **employed productively**, then the income can be regarded as "**earned**". Could we say the same for return on services such as loans for consumer credits or revenues obtained on bonds or bills?

Before lending money to customers, financial institutions often demand, as a prerequisite, either collateral or a guarantor in order to secure the return of the loan with interest. The idea behind is to minimize or rather to eliminate the likely risks of default. However, although quite rational from the point of view of lender, such prerequisites are against the very spirit and nature of market transactions, where the firms earn income in form of profit in return of the risks assumed. All attempts to minimize or eliminate such risks can be classified as both, unethical and unfair. Therefore, it would be rather difficult to justify such incomes as **earned**.

Interest on Loans for Investment

Investment is a necessary and crucial activity for the prosperity of any economy. It can be an output increasing investment with the given technology, thus increasing employment and total output, thus total wealth. Or, it can be in new product or process technologies, thus sustaining long term economic and prosperity growth. Some investments require a relatively low amount of capital to initiate production. But some, especially using high technology, output requires large, even huge sums of capital, i.e., savings productively employed.

Regardless of the scale of production, the entrepreneur may be in need of "external" financial funds to finance the new investment. In other words, the capital of the investor may be insufficient to initiate or expand production. The lacking capital may come from other people with savings and a willingness to share the risks, which make them co-partners. But it is not always easy to find such capital. Not infrequently, there are entrepreneurs in need of capital, on the one hand, and financial savings accumulated on the other. Financial institutions (**FIs**) may fill in the gap, and may loan money for investment to entrepreneurs. The action of the **FI** is a service transferring the "unproductive" savings to investors for "**productive**" employment. By doing this, the **FIs** employ labor(-er) as well as assume the risks of loan default. The outcome of such services supplied is the return normally called interest. What the **FI** does is economically rational and morally justified, as the **FI's** transaction contributes to prosperity. Thus, the income obtained on the services supplied is in fact an entrepreneurial profit and it is "**earned**" income.

Interest on loan for consumption

Providing consumer credit is one of the widespread transactions of the financial system. Such loans, if rationally used, often make a positive impact on the output of goods and services by increasing the current level of effective demand, which implies not only growth and increased wealth, but also more employment. There may be several reasons for seeking loans ranging from meeting urgent basic demands to the satisfaction of personal desires. Borrowings for the present enjoyment of desires, which can be postponed to a future date, constitute the major part of consumer loans. Purchasing a "new" car or a "new" house seem to give more satisfaction now, than in the future, for many consumers. And in return of this satisfaction, the credit user is prepared to pay some interest on loan.

If both sides, the credit supplier and the credit user, are satisfied with the transaction, than the credit system and the ensuing interest within the justifiable limits would seems to be both rational and justifiable. In fact, it would promote economic prosperity. The consumer enjoys the early satisfaction of a desire and pays interest, which is a personal choice. The **FI** employs labor and assumes the risk of default on re-payment. Therefore, the interest obtained by the **FI** seems to be **earned** income, or rather entrepreneurial profit.

If the person or family lacks sufficient income or assets to meet urgent and/or "basic" needs for decent living and seek loans, then we have a totally different situation at hand. There can be no economic or moral justification for demanding interest from people in such situations. In fact, no civilized society should leave their co-inhabitants in such distress and ignore their problems. If any money is loaned to such persons, technically, the interest may be regarded as entrepreneurial earning, but it would not be morally justifiable.

Interest on State or Company Debt Assets

State bonds or securities: It is a well-known global fact that practically all countries, developed or developing, have public debts of various sizes sometimes exceeding the GDP debtor nations. This situation is, in some cases, due to internal factors, such as mismanagement of the economy caused by the excessive and extravagant spending of public resources, and sometimes due to external events such as reduced demand for exports, which inevitably distorts macroeconomic balances. As a result, the state budget requires additional financial resources to keep up with the programmed performance and to sustain macroeconomic stability. Governments are normally faced with two options, under the circumstances;

1. printing money;
2. borrowing.

According to the dominant doctrine, printing money is an obsolete method not advised by financial experts and institutions such as the IMF. Borrowing has become the globally resorted to measure to overcome financial shortages, which is not free of various shortcomings and imperfections. The extent and nature of the borrowing can and often does, become a burden to the detriment of the fiscal performance and the further development of an economy. For instance, the rate of interest on borrowed money still exerts a heavy burden on the Turkish economy and the budget.

The rate of interest on loans varies from country to country depending on the economic credibility of the individual nations. But it is generally higher for developing countries (**LDCs**) on account of the so called "higher risks". Borrowing can be a relief of immediate burdens but in the medium and long term, the repayment of loan and the interest rate often seem to be a heavy burden for a long period of time which retards the development efforts in **LDCs**. It is an established fact that the income in the form of interest not infrequently exceeds the profits (earned income) on investments.

The first and best policy for the government seems to be maintaining a "balanced budget" free of fiscal deficits, thus avoiding public debts. Though highly desirable, balancing state revenues and expenditures cannot always be attained and unanticipated deficits may arise. As a result, countries may end up in a situation, which inevitably compels the governments to borrow to finance the deficit.

The first-best remedy seems to be taking some precautionary measures to overcome unanticipated fiscal deficits, such as setting aside some funds to be used in times of need. Such precautionary reserves would provide not only funds for a readjustment process to a new situation without fiscal problems but also provide time for decision-makers to take the necessary precautions. However, the global data indicates that the countries are far away from keeping and maintaining such reserves.

The second-best policy, according to the prevailing dominant doctrine, in case of unanticipated fiscal problems without reserves, seems to be borrowing. As mentioned above, probably all countries around the world have public debts, which inevitably lead to fiscal and other constraints, as we observe in almost every corner of the world. The constraints arising from a debt burden are, in most cases, to the detriment of development efforts, especially in **LDCs**. Even the interest rate on the loans of international organizations like the IMF is often quite above the international market rates, although the officially declared aim is to help the nations.

In theory, the loans to states (governments) to finance budget deficits are subject to the risks of deference and bankruptcy. Bankruptcy is, in the case of nations, out of the question. A nation does not go bankrupt. However deference on repayment of loans and/or interest is not a rare case. The lenders would naturally expect some reward, e.g., interest. Because of the risk assumed and the labor efforts employed in supplying the debt service, the income of the financial institution can be regarded as entrepreneurial "profit", and therefore can be classified as **earned** income.

Company Bonds and Bills

To set up a new plant, to expand production, or to develop a new technology, an enterprise might be in need of some financial funds (capital). One of the options of the producing firm is to borrow from **FIs**. The second option is to export debt assets in form of short-term obligations (bills) or long-term bonds. In both cases the enterprise has to pay interest on the loan. And in both cases, the interest obtained by money lenders is **earned** income, as they assume some risks related to output.

Instead of borrowing money and paying interest, enterprise could prefer to sell some shares. Since the new share-holders are directly assuming risks with supply of goods or services, their income will be classified as entrepreneurial profit, i.e., **earned** income.

Inflation and Interest

In the above analysis regarding interest, we implicitly assumed an inflation-free environment that is zero inflation. Now we shall see how inflation affects interest, or rather, how the inflation rate should affect the interest rate?

Case-1: Assume that a finance company lends 100 TL to a producer or a consumer with zero rate of interest for a year. Further assume that the inflation rate after a year is 10 percent. Inevitably, the finance company would suffer a loss of 10 percent in the real value of money in terms of purchasing-power.

Case-2: Assume that the finance company charges 10 percent interest rate on loan, which is exactly equal to the inflation rate at the end of period. The company will make no loss, but no profits, either.

The lesson to be drawn from the two cases with inflation is that compensation payment for inflation on borrowed money is not only economically rational but also ethical. Therefore, it is an economically rational behavior for a profit driven financial company to charge an interest rate above the expected inflation rate on the loaned money.

Final Remarks Concerning the Interest Rate

Two major negative economic impacts may occur when interest is paid on hoarded money. Firstly, the risk-taking entrepreneurs will have to pay a relatively higher rate of interest on their loans, as the financial intermediary will have to pay interest on the savings on the hoarded money. For instance, if the interest on savings is 10 percent, the risk-taking entrepreneur who wants to borrow funds for the supply of his products will have to pay interest at a rate above 10 percent in order to compensate the intermediary for the risks he has undertaken in loaning the money. If the interest rate on deposits was, say zero percent, the interest rate on loans for the supply of products would be lower. Accordingly, both the production costs and the sale price would be lower, cet. par.

Secondly, rewarding the deposits of hoarded money by giving interest would be likely to reduce the potential effective demand, because, at least some of the money holders would be inclined to make use of the interest obtained from their deposits to increase their hoard of money. Each unit of money not spent leads to a reduction in the potential demand for goods and services, thus reducing output and employment.

It is said that interest is the price of using the money (deposits). This is true if it refers to an economic transaction between the risk-taking money lender and the risk-taking entrepreneur, as both sides take risks. But, it is a rather different story in the case of obtaining a reward on the deposits. The money holders normally assume no risk of a default on the loan, but the financial intermediary does. In other words, it is the financial intermediary who would suffer a loss if the borrower cannot pay back the loan, not the deposit holder. What could the economic rationality of this type of interest be, if there are no risks and no output of any kind?

In the present financial system, the money-savings make more money without making any contribution to the supply of goods or services. This situation cannot be defended as economically rational or ethical just or fair. An economically rational financial system favoring the individual as well as of the whole society should encourage and promote the **productive employment** of hoarded money towards supplying goods and services.

4- Rent: "Unearned Income"

Rent is another kind of income subject to everlasting moral as well as economic arguments. It is an income **"not obtained by the efforts of labor but simply by appropriation"**. How the property came into possession of the owner in the first place is a rather important issue from the point of income distribution, but is beyond the scope of this work. Three kinds of rent will be analyzed below.

1. One kind of rent is related to the "excessive" profits above the market average arising from **market imperfections** such as macro-economic policies, black-markets, cartels, oligopoly, monopoly. End-users have to pay higher than normal prices, which increases the profits above the normal profit-level without any additional costs or efforts. Since this type income does not involve any labor effort or competitive risks, it can be categorized as **"unearned"** income.

2. Another type of income categorized as rent is related to the ownership of property, such as land, building, flats, etc. For instance, the owner of a piece of land may obtain income, as a result of efforts of those who actually toil the land. Income accruing may be in the form of money or "in kind", like crop-sharing. Or, in a similar fashion, the owner of a building or a flat may let out the owned property for a pre-determined amount of money for a given period and receive an income called "rent". Such types of incomes can be categorized as **earned** income, provided that the property is, in the first place, acquired through appropriate means.

   Assume that the rental price of hired land or building or flat is 100 TL and for some reason other than inflation the rent increases to 150 TL, say due to an unexpected rise in demand. Is the 50 TL increase in income received by the property owner **earned** or **unearned?** The answer depends on how one subjectively evaluates the income.

3. A third kind of rent arises as a result of "Ricardian" style differences in the land productivities of different natural qualities. Assume that there two equal size pieces of land with different qualities. The same quality and amount of inputs are used to grow tomatoes in both lands. But the first land is assumed to be 20 percent more fertile (productive) per unit, which implies a higher level of income and profit rate for the owner. Since the difference in income is not due to market imperfections, the income can be categorized as **earned**.

The differences in income from rented buildings or flats of equal size but in different locations can be treated similarly. For instances, the rental price of a flat in a suburb is normally lower than a flat of equal size in the town-center. There seems to be no economic or logical ground to classify such rental income differences as **unearned**.

Functional Income Distribution

Unequal distribution of income has always been one of the major problem areas and often a cause of embarrassment for both economic science and economists. The Classical economic analysis is used to emphasize functional income distribution among the three classes, i.e., workers, capitalists and landowners. "Modern" economic textbooks of a "neoclassical" heritage analyze functional income distribution in terms of "payments to the two factors of production", i.e., capital and labor, in accordance with their "marginal productivities". Contemporary researchers on income distribution seem to place more emphasis on an individual, group and/or cross-country income distribution analysis, rather than on functional distribution.

As we have seen, the income to be distributed is generated by the **productive employment** of the resources necessary to meet the needs and desires of end-users. There were a number of **production factors** but only two **productive factors** i.e., nature and labor. The types of incomes and the income receiving factors, on the other hand, were more than two:

3. Labor **earning** wage.
4. Capital (savings productively employed) **earning** a profit.
5. **Return** on unproductive savings; and
6. All the different kinds of rent.

The focus of this study concerns the functional income distribution between two groups the capitalists and the workers (laborers). The reason for looking at the income distribution of these two groups is that the generation of added value only occurs when "capital" is used together with laborers to transform nature's products.

In the following part, functional income distribution will be studied under two subcategories:
1. With a "given" technology.
2. With technological progress.

1- Income Distribution with a "Given" Technology

Assume that the following hypothetical figures represent the initial production relationships in a closed economy with a "given" technology and no government interference. **w** denotes wages, **TC** cost of production, **TR** total revenue, π profits, **r** rate of profit and **LWC** total wage cost:

$w = 100$ TL
$L = 500$
$p = 18$ TL
$q = 10,000$ pieces
$LWC = w * L = 100 * 500 = 50,000$ TL
$OC = FC + VC = 80,000$ TL
$TC = LWC + OC = 130,000$ TL
$TR = 18 * 10,000 = 180,000$ TL
$\pi = TR_t - TC_t = 50,000$ TL
$r = \pi / TC = \sim 38$ percent
$VA = \pi + LWC = 50,000 + 50,000 = 100,000$ TL

Distribution of hypothetical income:

π / VA	=	50 percent	**share of profits**
LWC /VA	=	50 percent	**share of labor**

In fact, the distribution of income has never been equal. But, for the sake of argument, we assume it is.

Wage Rise and Income Distribution

Equal distribution of income can change in two ways;
1. by changing price level, and thus the profit rate; or
2. by changing the wage rate.

A wage rise would reduce the profit's share while increasing that of labor, which demonstrates that the interests of the income sharing groups are in opposition to each other. To the extent that the labor succeeds in raising the wage rate with a given income level, there will be an improvement in the labor's share at the expense of the capital-owners.

Assume that there is a wage rise by 20 percent, cet. par.

$\Delta w = 20$ TL

And the new wage would be;

$w_{t+1} = 120$ TL

As a result, both the labor wage costs as well as total costs will increase.

$LWC_{t+1} = w_{t+1} * L_{t+1} = 120 * 500 = 60,000$ TL

$TC_{t+1} = LWC_{t+1} + OC_{t+1} = 140,000$ TL

The change in costs would change all the other outcomes.

$\pi_{t+1} = TR_{t+1} - TC_{t+1} = 40,000$ TL

$r_{t+1} = \pi_{t+1} / TC_{t+1} = \sim 28$ percent

$VA_{t+1} = \pi_{t+1} + LWC_{t+1} = 40,000+60,000 = 100,000$ TL

New distribution of income would clearly improve labor's share:

π_{t+1} / VA_{t+1} = 40 percent **share of profits**

LWC_{t+1} / VA_{t+1} = 60 percent **share of labor**

Although the total income (**VA**) has not changed (100,000 TL), the share of profits fell from 50 percent to about 40 percent while that of wages increased from 50 percent to 60 percent. An increase in the profit rate would make the opposite impact on income distribution.

2- Income Distribution with "Technological Progress"

An important aspect related to the mental labor services would be the concept of technological progress, which is considered to be a constant source of additional income generation. Technology can be defined, in a narrow sense of the meaning, as the "knowledge" materialized in commodities, which is in fact a product of mental labor services, e.g. human capital (Gürak, 2006).

There are two major reasons for investing in "new" technology:
1. Not to lag behind competitors, in fact, go ahead of them if possible.
2. To maximize long term profits.

And there are two types of technological change that serves these purposes:
1. A new production method for given product.
2. New products and/or new production methods.

Let us analyze how these technological changes affect income distribution.

2-a: New production method for given product:

In order to survive in the long term in a competitive environment, the firms have no choice but search for "new production methods" of "given" products, in order to reduce per unit production costs. Assume that among the initial economic figures only the quantity has changed as a result of the new technology and increases from 10,000 to 12,000 pieces, a rise of 20 percent.

$$w = 100 \text{ TL}$$
$$L = 500$$
$$\mathbf{p} = 18 \text{ TL}$$
$$\mathbf{q} = 12{,}000 \text{ pieces} \qquad \Delta\mathbf{q.} = 2{,}000 \text{ pieces}$$
$$\mathbf{LWC} = \mathbf{w} * \mathbf{L} = 100 * 500 = 50{,}000 \text{ TL}$$
$$\mathbf{OC} = \mathbf{FC} + \mathbf{VC} = 80{,}000 \text{ TL}$$
$$\mathbf{TC} = \mathbf{LWC} + \mathbf{OC} = 130{,}000 \text{ TL}$$
$$\mathbf{TR} = 18 * 12{,}000 = 216{,}000 \text{ TL}$$
$$\boldsymbol{\pi} = \mathbf{TR_t} - \mathbf{TC_t} = 216{,}000 - 130{,}000 = 86{,}000 \text{ TL}$$
$$\mathbf{r} = \pi / \mathrm{TC} = \sim 64 \text{ percent}$$
$$\mathbf{VA} = \boldsymbol{\pi} + \mathbf{LWC} = 86{,}000 + 50{,}000 = 136{,}000 \text{ TL}$$

As a result of the technological change, the total income generated increased by 36,000 TL to 216,000 and the distribution of income changed in favor of the capital-owner, even though the real wage and the total labor income has not changed:

$$\pi / \mathrm{VA} = \qquad \sim 63 \text{ percent} \qquad \textbf{share of profits}$$
$$\mathrm{LWC} / \mathrm{VA} = \qquad \sim 37 \text{ percent} \qquad \textbf{share of labor}$$

2-b: New products and/or new production methods:

Introducing "new" technologies to produce "new" products usually accompanied by "new" production methods in turn implies new and higher profit opportunities than the average; at least this is the expectation behind the introduction of the new technologies. If the "expected" profit rate was not higher than the average or there is shortage in competition, there would be insufficient incentives to introduce "new products".

In general, new products lead to a larger income with new and higher profit opportunities and probably a larger share of the total income for the capital owners, that is until the next round of wage rise negotiations, of course. What the rate of profit or income distribution would be is uncertain. For, there are no previous quantities to compare with.

Should all the income benefits of a "new" technology accrue solely to profits? Who are entitled to accrue the benefits of the productive knowledge (technologies) originating from the mental labor services based on thousands of years of accumulation and improvements?

The inventor? The enterprise? Mankind? Or, all three?

To the extent that an enterprise finances the invention and/or the innovation of a new technology, and assumes the related risks; it is economically rational and morally justified to reap, at least, some of the accruing benefits such as profits. But, some of the benefits should belong to the inventor(s), who produce the new technology, some to the community from which inputs like education, infrastructure, etc., are provided as well as to mankind, which provided the common "infrastructure and environment" shaped over thousands of years. Restructuring the "ownership rights on intellectual property" facilitating an easier access to productive knowledge by the firms of the less developed countries and eliminating imperfections in the technology market could make the world a better place to live in.

Long-Term Wage Rise and Income Distribution

As seen above, given the wage rate in the short term, technological innovations, normally, tend to increase the rate of profits and the share of these profits, in regard to total income. Therefore, all commercial firms constantly aspire to employ more productive technologies partly in order to survive the competition and partly to realize in the long term, the maximum profits possible. If we assume that all inputs, including laborer, are efficiently used, each cost-reducing productivity growth due to the new production method[16] will lead to, in the short term, a deterioration in the distribution of the functional income for the wage-earners, even if the wage level remains the same.

$$w_{t+1} = w_t$$

But;

$$VA_{t+1} > VA_t$$
$$r_{t+1} > r_t$$
$$\pi_{t+1} > \pi_t$$
$$\pi_{t+1} / VA_{t+1} > \pi_t / VA_t \qquad \textbf{change in capital's share of income}$$
$$w_{t+1} / VA_{t+1} < w_t / VA_t \qquad \textbf{change in labor's share of income}$$

The "new" technology generates more income but also shifts the relative shares of the income generated, which deteriorates for the employees and improves for capital-owners. It is important to note that there is no change in the real wages. To put it differently, the "new" technology makes the capital-owners richer while the position of employees' remains unchanged with a "given" wage level.

---

16   The type of new technology assumed refers to a "new" production method of a "given" product.

Laborers respond to this new situation by demanding a wage rise in the next round of negotiations with the capital owners, as stated above. The outcome depends partly on the economic conditions prevailing at the time and partly on the abilities and strength of the negotiating parties.

Concluding Remarks

As demonstrated in this study, production factors and productive factors are not the same thing. There are only two productive factors but there may be thousands of production factors. Therefore, given the gifts of nature, there is only one productive factor adding value to products; that is the laborer. Capital is absolutely not a productive factor capable of producing or adding value to products as assumed by the neoclassical ideologist. Capital is a produced production factor indirectly contributing to the production of any added value.

Income is generated and/or increased by the productive employment of savings combining the inputs of production, including the labor services, to meet the wants and needs of end-users in return for profits. The laborer, the only value-adding factor, earned wages in return of the labor services they supplied. Without the supply of the productive knowledge of the labor services, mankind would still be living and attempting to survive in a jungle-like environment.

The income generated should be distributed among two economic groups; wage-earners and the risk assuming capital-owners. But other economic agents are receiving a share of this income in the form of interest and rent without actually making any direct contribution to the generation of this income.

Interest on loans (savings) for investment and the satisfaction of "desires" may be considered rational and justifiable. But interest on loans to meet the "basic needs" of life cannot be defended as economically rational in a modern society. Nor can it be morally justified.

As to the generation and distribution of income, there is a direct relationship between the technological advancement of a country and the level of total income. The higher the technological level, the greater the expected total income will be. The Neoclassical assertion that capital has marginal productivity and functional income is distributed according to the marginal productivities of two factors, labor (L) and capital (K) is nothing but an ideological fallacy. It is common knowledge in regard to the economic relations in the actual world that income before tax is determined, in general, by the bargaining power of the labor and employer unions.

54

In regard to income distribution, the study demonstrated that:

- As the productivity grows, the real wages remain the same while their relative share in income declines, in the short term.
- Meanwhile the relative share of profit increases.
- As a result, productivity growth due to a "new" technology decreases the income distribution for labor, in the short term. The long term position of labor is up to their bargaining strength at wage negotiations with the firms, as well as the prevailing conditions in the markets.

References

Blanchard, O.   2003 **Macroeconomics, 3. Ed.**
                            Prentice Hall
Blaug, M.   1980 **The Methodology Of Economics.**
                            Cambridge Uni. Press, Cambridge.
                      1990 **The History of Economic Thought.**
                            Edward Elgar Pub.. Ltd., Hants.
Bohman, R.S.   1990 Smith, Mill & Marshall On Human Capital
                            Formation. *History Of Pol. Economy*, Vol. 22:2
Böhm-Bawerk, E.   1896 "The Positive Theory of Capital and its Critics"
                            *The Quarterly Journal of Ec;.*Vol.10; No. 2
Caldwell, B.J.   1994 Beyond Positivism: Economic Methodology In
                            20. Century , Routledge, London.
Clark, J.B.   1894 **"The Genesis of Capital".**
                            Publications of AEA, Vol.9; No.1; 64-68
Cohen, A.J. –
Harcourt, G.C.   2003 Whatever Happened to the Cambridge
                            Capital Theory Controversies?
                            Journal of Economic Perspectives,Vol.17:1
Dougherty, C.   1980 **Interest and Profit.**
                            Methuen & Co. Ltd., London.
Edgeworth, F.Y.   1894 "Prof. Böhm-Bawerk on the Ultimate
                            Standard of Value",
                            The Economic Journal, Vol.4; No.15.
-----"-----   1894 "One More Word on the Ultimate
                            Standard of Value",
                            The Economic Journal, Vol.4; No.16, Dec.
Fetter, F.A.   1977 **Capital, Interest and Rent.**
                            Essays in the Theory of Distribution.
                            Sheed Andrews & McMeel, Inc., Kansas City.
Friedman, M.   1976 **Price Theory.**
                            Aldine Publ. Co., Chicago.
Gürak, H.   1999 On Productivity Growth
                            YK-Economic Review, Dec,Vol.10, No:2.
--- " ---   2000-a Economic Growth and Productive Knowledge
                            *YK-Economic Review,*June, Vol.11, No:1.

--- " ---   2000-b Verimlilik Artışları (Productivity Growth)
                         Verimlilik Dergisi, Eylül-Ekim, MPM, Ankara.
--- " ---   2003 "Hidden Costs of Technology Transfer"
                         YK-Economic Review; June; Istanbul.
--- " ---   2004 On Value and Price
                         YK-Economic Review, June,Vol.16,No:1.
--- " ---   2006 Ekonomik Büyüme ve Küresel Ekonomi
                         Ekin Kitabevi, Bursa.
Harrod, R.F.   1937 "Mr. Keynes and Traditional Theory".
                         Econometrica, Vol.5; No.1, Jan. 74-86
Harris, D.J.   1978 **Capital Accumulation & Income Distribution.**
                         Routledge & Kegan Paul, London.
Hasan, Z.   2008 **Theory of Profit from Islamic Perspective**
                         http://mpra.ub.uni-muenchen.de/8129
Hausman, D.M.   1981 **Capital, Profits and Prices.**
                         Columbia Uni. Press, New York.
Hayek, F.A.   1984 **Money, Capital and Fluctuations.**
                         Routledge & Kegan Paul, London.
Hicks, J.   1965 **Capital And Growth.**
                         Oxford Uni. Press, London.
--- " ---   1979 **Causality In Economics.**
                         Basil Blackwell, Oxford.
--- " ---   1983 **Classics and Moderns.**
                         Basil Blackwell Publ., Oxford.
Howard, M.   1983 **Profits in Economic Theory.**
                         The Macmillan Press Ltd.
Kaldor, N.   1938 **"On The Theory Of Capital: A Rejoinder"**
                         Econometrica, Vol.6; No.2, April, 163-176
-----"-----   1939 "Capital Intensity and the Trade Cycle"
                         Economica, Vol.6; No.21, Febr.; 40-66
-----"-----   1960 **Essays On Value And Distribution.**
                         G.Duckworth & Co. Ltd., London.
-----"-----   1989 Further Essays On Economic Theory & Policy
                         Duckworth, London.
Kalecki, M.   1987 Selected Essays On The Dynamics Of
                         Capitalist Economy. Cambridge Uni. Press.
Kasper, W. –
Streit, M.E.   1998 Institut**ional Economics**
                         Edward Elgar Publ.

Knight, F.H.	1935	"Prof. Hayek and the Theory of Investment"
		The Ec. Journal, Vol.45; No.177; March; 77-94
-----"-----	1964	**Risk, Uncertainty And Profit.**
		Sentry Press, New York.
Lucas, R.	1988	"On The Mechanics Of Economic Development"
		Journal Of Monetary Ec., July, 1988,342
Mac Vane, S.M.	1893	**"The Austrian Theory of Value".**
		Annals of the American Academy of Political and Social Science, Vol.4; Nov. 12-41
Machlup, F.	1935	"Prof. Knight and the 'Period of Production'"
		The Journal of Pol.Ec., Vol.43; No.5; 577-624
Marx, K.	1970	**Ekonomi Politiğin Eleştirisine Katkı**
		Öncü Kitabevi, İstanbul.
-----"-----	1975	**Ücret, Fiyat ve Kâr**
		Sol Yayınları
-----"-----	1976	**Capital, Vol. I**
		Penguin Books.
-----"-----	1977	**Capital, Vol. II**
		Lawrance & Wishart, London.
-----"-----	1981	**Capital. Vol. III**
		Penguin Books.
Marshall, A.	1961	**Principles Of Economics, Vol. I & II**.
		Macmillan And Co., London.
Meek, R.	1973	**Studies in the Labor Theory of Value: From Smith to Ricardo.**
		Lawrance and Wishart, London, 2nd Ed.
Mill, J.S.	1986	**Faydacılık (Utilitarianism)**
		Milli Eğitim Basımevi, İstanbul.
Pasinetti, L.L.	1979	**Lectures on the Theory of Production**.
		The Mcmillan Press Ltd., London.
Ricardo, D.	1990	On The Principles Of Political Economy And Taxation. Cambridge University Press.
Robinson, J.	1962	Economic Philosophy.
		Penguin Books.
-----"-----	1966	The Accumulation Of Capital.
		Macmillan, London.
-----"-----	1972	"The Second Crisis of Economic Theory"
		The American Ec. Review, Vol.62; No.1/2; 1-10
Romer, P. M.	1990	"Endogenous Technological Change"
		Journal Of Political Economy, Vol.98, Oct.

Samuelson, P.A.   1967 **Economics**.
                         McGraw-Hill Book Co., New York.
Schumpeter, J.A.   1954 **History of Economic Analysis**.
                         Oxford Uni. Press, New York.
--- " ---   1959 **The Theory Of Economic Development**
                         Harvard Uni. Press, Cambridge.
-----"-----   1970 **Capitalism, Socialism And Democracy**.
                         Unwin University Books, London.
Smith, A.   1976 An Inquiry Into The Nature And Causes
                         Of The Wealth Of Nations, Vol. I & II.
                         Liberty Classics, Indiana
Smithies, A.   1935 "The Austrian Theory of Capital in Relation to
                         Partial Equilibrium Theory", *The Quarterly
                         Journal of Economics*, Vol.50; No.1; 117-150
Solow, R.M.   1957 "Technical Change and the Aggregate Production
                         Function", **Rev. of Ec. and Statistics**, August.
Sowell, T.   1974 **Classical Economics Reconsidered**.
                         Princeton Uni. Press, Princeton, New Jersey.
Sraffa, P.   1977 Production Of Commodities By Means Of
                         Commodities. Cambridge Uni. Press, Cambridge.
Stanfield, J.R.   1979 **Economic Thought and Social Change**.
                         Southern Illinois Uni. Press.
Ulutan, B.   1978 **İktisadi Doktrinler Tarihi**.
                         Ötüken Neşriyat, İstanbul.
Weintraub, S.   1977 Ed. **Modern Economic Thought**.
                         University Of Pennsylvania Press.
Wieser, F.   1891 **"The Austrian School & the Theory of Value"**
                         The Ec. Journal, Vol.1; No.1; March, 108-121
Young, T.J.   1978 **Classical Theories of Value: From Smith to
                         Sraffa**. Westview Press, Colorado.

2- ON VALUE AND PRICE

An alternative approach to value & price theory

> "Where is the discussion of mental labor and technological change in the theory of value and price?"
>
> H. Gürak

Introduction

Only lip service was paid to the "growth theory" until the 1950's. Since the 1950s, however, concepts like technological change and mental labor (human capital) have been re-discovered as a vital, essential and indispensable ingredient of the growth theory. Nowadays there is wide a range of endogenous growth theories constructed on the back of these re-discovered concepts. In spite of many shortcomings the trend is promising and may enable modern economists to construct more realistic and reliable growth models capable of accounting for actual global economic facts and developments.

Meanwhile, however, the backbone or the basis of all these theories, i.e., the value and price theory still fails to keep pace with developments in growth theory. It neglects to incorporate key concepts such as technological change and mental labor in its analysis. As a result, the "modern" but sterile price theory is bound to fail to properly account for the price formation in the "real" markets. Nor does it provide any appropriate or sound premises for the construction of any related economic theories that are affected or influenced by the price signals coming from the real markets.

The Hypothesis

The hypothesis of this paper is to show that all value-created or added to the natural resources originates from the labor-power that offers two kinds of services; **mental labor** and **physical labor.** The former, the mental labor, is the original source that constantly introduces "new ideas" that is "new technologies" to transform natural products. Meanwhile the latter, physical labor plays a complementary role in accordance with instructions delivered from the brain. As the paper aims to analyze values and/or prices, all concepts and definitions refer to an exchange-economy in which, the natural resources are accepted as a given

61

and **mental labor is acceded as the genesis and incessant source of all the value created**.

The approach is basically a labor embodied approach, but somewhat distinct from the Classical approaches, including those of Smith, Ricardo and Marx. Though it acknowledges the labor-power as the genesis and a constant source of added value, it makes no claim to be an "invariable" measure nor does it claim that the profit (surplus value) is an "unpaid" or a "surplus" part of labor. In addition, it does not make any claim that proper exchange relations should be based on "equal quantities" of the labor-time employed.

Given demand, new exchange relations are determined by the new conditions created by "new technologies", which are a result of the **productive process of the human mind**, i.e., creative mental labor, cet. par. In other words, the productive faculty of the human mind is assigned a key role in all the exchange relations of relative values and price formation. As Marshall pointed out:

> Man cannot create material things. ... indeed he may produce new ideas ... his efforts and sacrifices result in changing the form or arrangement of matter to adapt it better for the satisfaction of wants. All that he can do in the physical world is either to re-adjust matter so to make it more useful, ... or to put it in the way of being made more useful by nature... (Marshall; 1990; p.53)

The emphasis of this analysis is on the "relative" values and prices followed by a brief analysis of the actual market values and prices. The reason for that is not because relative exchange relations provide a better premise for analysis of "actual" price formation but because it has been customary to start with relative values since the time of the Classical economists.

Why Price Theory?

The value or price theory holds a very crucial and central position in economics and has long been considered as the basic tool and as the "essence" of all economic analysis including economic predictions. As is well known, producers as well as consumers, adjust their market behavior according to the price signals, which determine the allocation of their resources. Price signals are capable of influencing crucial variables like the growth rate, inflation, employment, etc. Therefore, it is imperative to have access to a competent price theory, which forms a logical, consistent as well as a reliable basis for all actual transactions. As the "modern" theory fails to satisfy the last condition, a need for an alternative theory arises.

A realistic price theory should not only be capable of explaining exchange ratios, e.g., the relative prices, between two commodities, but also the market prices of all the commodities (tangibles) produced. In addition, this price theory should also be able to explain the pricing system in the intangible service sector, which appears as the neglected "stepchild" of present price theories. Nowadays, the service sector in terms of output and employment accounts for the greater part of the GDP in modern economies and displays features distinct from the manufacturing sector producing tangible goods.

And, perhaps most important of all, this price theory must be able to account for the past, present and future source of all value generation, the transformation of these values into prices and the distribution of functional income between wages, profits and interest. Only then could one have a more realistic insight into, and a clear interpretation of actual economic relations. Such a theory should also pave the way for the further development of both sound and realistic theories in the related fields like growth, trade, employment, etc.

Towards a New Mindset"

As Schumpeter quite rightly pointed out:

> "....... in practice we mostly do not start from a vision of our own but from the work of our predecessors or from ideas that float in the public mind." (Schumpeter, 1954; p. 562)

Throughout this study, the purpose is to escape from the old habituated mindset (thoughts and expressions), which naturally shapes or heavily influences the mindset of any student of economics. This is not an easy task after years or decades of indoctrination. As Keynes, put it:

> "The difficulty lies, not in the new ideas, but in escaping from the old ones, which ramify, for those brought up as most of us have been, into every corner of our minds." (Keynes, 1973; Preface, p. xxiii)

In the following sections we will first analyze exchange relations where only physical labor-power is employed in a simple economy. Then, the analysis will concentrate on exchange relations where "creative mental labor" is incorporated into the model of a simple economy, followed by an attempt to convert values into prices.

A Brief Historical Review

In contrast to the "modern" approach, economic science before 1870s was treated more like an interrelated social science. Inexact but actual, rather than exact and fictitious economic relations were the point of departure. Theory of value was considered as the backbone of the political economy and the concepts like "justice" and "equality" were not regarded as irrelevant. Extremely abstract mathematical reasoning applied to economic relations, free of any kind of "human weakness" is the result of more than a century of attempts to make economic science an "exact" science like astronomy and physics and economic scholars have come a long way in this respect. British economist Jevons once proudly claimed that his model of exchange relations did;

> "... not differ in general character from those which are really treated in many branches of physical science". (Blaug, 1990,p.147)

But the models created were nothing but oversimplified idealizations, a hypothetical version of reality, a "virtual economy". Marshall had foreseen this pathetic trend and warned economists to be cautious in their application of mathematical models and not to transform economic science into a branch of mathematics, but he did not succeed. In time, physical sciences bowed to the developments in science and underwent drastic changes upgrading itself in accordance with the new discoveries arising from Newtonian physics and "Quantum" physics, while economics, as a "downgraded" natural science", has remained faithful to its obsolete a priori philosophical criteria.

For many prominent economists, the neoclassical heritage still represents the **"holy grail"** of any analysis capable of revealing the true nature of economic man and his actions. Any dissent from this holy fount of eternal truth is regarded as a grievous error, if not a "mortal" sin. An outside observer can easily get the impression that the neoclassical teaching is the **final frontier** and the **highest stage** of economic thought. Naturally, all scholars do not agree. As Hicks put it:

> "Pure economics has a remarkable way of producing beavers out of hats -apparently a priori propositions which apparently refer to reality. It is fascinating to try to discover how they got in; for those of us who do not believe in magic must be convinced that they got in somehow." (Hicks, 1983,p.367)

Hicks continues:

> "Economics is a social study. It is concerned with the operations of human beings, who are not omniscient, and not wholly rational; who (perhaps because they are not wholly rational) have diverse, and not wholly consistent, ends. As such, it cannot be reduced to a pure technics." (Hicks, 1983,p.289)

The critical question is; do we have a logical and consistent alternative theory?

Unfortunately, there seems to be no positive reply to this question, yet. The number of academic units and academicians with "heterodox" orientation has been growing in numbers recently but, there seems to be no widely acknowledged consistent and logical alternative theory, so far. In this section, an attempt will be made to fill this gap based on a different approach based on a distinctive "labor-value" theory which is the basis of all economic theories. Let us begin with the "original" sources of value.

The Original Sources of Value: Nature and Labor-Power

Let us begin with a definition of "value". In economic terminology, the "value" of a product whether it is a commodity or a service, is the "relative worth" that can either be exchanged for other product(s) (exchange-value) or be used for personal consumption (use-value). The latter is a subjective concept and its magnitude depends, given income, on the ranking of products in accordance with subjective needs and preferences. The former, "exchange-value", expressed in terms of market prices, depends partly on the costs of production and partly on the demand for the specific product, given the competitive environment.

There are only two initial sources of value-generation, the labor-power and nature. Nature provides a wide range of exogenously given and unlabored (unprocessed) objects with use-values. The labor-power adds value to the nature's objects by transforming (reshaping) them, into commercial products containing exchange-values, by utilizing human mental and physical capabilities. In other words, the past and present services of labor-power transform natural products into useful products either for immediate consumption or into inputs of production or into capital goods which in turn are used to increase the productivity of the labor-power.

In a modern society, to initiate a value creation process, the capital owner, e.g., the entrepreneur has to have access to money or capital (savings) to combine (purchase or hire) the material inputs of production such as raw materials, machinery, tools, energy, etc., using the services of labor-power, e.g., mental and physical labor. This feature of savings might give the impression that "capital" is one of the "productive" factors of production, although there is in fact no universally acknowledged definition of "capital" (Hausman, 1981). In some analyses it appears in monetary form and in others as physical inputs like tools, machinery and sometimes it even refers to both. Both, money-capital and capital goods are the necessary ingredients of production but are certainly not productive ones in the sense that the value adding labor-power or the initial supplier of objects, i.e., nature, are. At first glance, savings seem like a "productive" factor of production as it gives rise to the employment of productive labor-power along with the implements of production. But, money as such, e.g., savings, cannot be productive, as it is not capable of producing any value. Therefore, money is definitely not one of the "productive" factors.

Capital goods are just as productive as savings. Capital goods being the man-made inputs of production help to increase the productivity of the laborer. However, there would be no capital goods unless natural objects were transformed by the laborers. Capital goods can only transfer value to the product at the rate of its depreciation; nothing more, nothing less.

The Labor-power

The labor-power is embodied in products in the form of mental and physical labor, e.g., skilled and unskilled labor. Thus, the labor-power can be defined as;

> "... the aggregate of those mental and physical capabilities existing in the physical form, the living personality, of a human being, capabilities which he sets in motion whenever he produces a use-value of any kind". (Marx, Vol. I: p. 270)

It is the mental component of labor that generates the productive knowledge that accounts for both the quantitative growth and qualitative improvement of physical objects as well as of the services supplied. Physical labor is a necessary ingredient of production but not a sufficient one to increase the added value of nature's produce on its own. Without the contribution of mental labor, it would not be possible to produce the sophisticated goods and services and reach the contemporary standards of living that, some of us all around the globe, so lavishly enjoy. In other words, to possess an exchange value, the contribution of both mental and physical labor, to a more or lesser degree, is imperative at every stage of production. But, while physical labor's contribution is rather limited, the mental labor is capable of continually creating new "added values".

The contributions of mental labor-power (new technologies) can be divided into two groups:

1. **New goods or services** (entirely new ones or old ones in a new form which are usually accompanied by "new" methods of production);
2. **Existing goods and/or services** but employing "new" production processes which reduces the cost per unit output

Value creation due to new technologies, as indicated above, can also be referred to as **macro-productivity** growth. But, there are, certainly, other methods of increasing the productivity with a "given" technology, which may be referred to as **micro-productivity** growth (see Gürak, 2000) which has limited impact in the short term until it reaches optimum levels.

To sum up; it is the mental ability of the labor-power that accounts for ever increasing value creation and sophisticated living standards. But it would have no significance if there were no gifts of nature to be transformed into useful things. Men and nature are, therefore, the two indispensable and inseparable sources, or **"complementary productive factors"**, of wealth. And all physical products, no matter how complex and sophisticated, can be considered as arising from to nature's gifts i.e., raw materials, if the input of past and present mental and physical labor content is discounted. Thus, every product may be reduced, in the final analysis, to nature and labor-power.

Mental Labor (Productive Knowledge) and Value Generation

The crucial and central question in relation to value or price formation is: What are the conditions determining the relative exchange-value of a product? Is it the supply-demand conditions? Is it the labor "embodied" in, or "commanded" by the final products? Or both?

Value Generation - A Simple Model

Below, a simple alternative labor embodied model of value-generation will be presented to see how the "productive knowledge" of the mind (mental labor) enters into the production process and how it effects the exchange relations in terms of relative values.

Let us begin, like Adam Smith did, with the well-known **"hunter model"** and assume two hunters and no tools of production at all, except for the services of labor-power with its two basic faculties, physical and mental labor. Being a quantitative concept, the "physical labor" is easily measurable by terms such as the hours worked, the days needed to complete a task or some other measurable unit while the latter term, the "mental labor", the source of the productive knowledge (new technologies), refers to a subjective concept and is unlikely to be estimated accurately.

Leaving aside, for the time being, this fundamental contribution of mental labor and let us assume that the two hunters in our model work 10 hours a day and the first one, Maria the deer hunter, acquires 2 deer a day while the second, Leyla a beaver hunter, acquires 4 beavers a day. If they had lived in a self-sufficient society meaning that all the catch is consumed within the family of each of the hunters respectively, there would be no need to engage in exchange relations. In the absence of exchange relations, there would be no exchange-values, either. But, our hunters do exchange.

Initial Exchange Conditions

Given tastes and preferences, assume that at the end of the day, the two hunters exchange one deer for two beavers, half a day's physical work, which is a fair exchange with respect to the physical labor expended, e.g., 10 hours' work, in each case. Leyla consumes one deer and two beavers; as does Maria. Nobody is better or worse off after the exchange, and the supply-demand is in balance after the egalitarian exchange.

Leyla's supply = 4 beavers = 10 hours' physical labor   (1)

Maria's supply = 2 deer   = 10 hours' physical labor   (2)

Total supply /per day = 2 deer + 4 beavers = 20 hours' physical labor. (3)

Leyla's consumption= 1 deer+2 beavers= 10 hours' physical labor   (4)

Maria's consumption= 1 deer+2 beavers=10 hours' physical labor   (5)

So far, our two hunters did not make any use of their mental faculties, e.g., mental labor (ML), in their daily work meaning there is no value-added except for the services of the physical labor (PL). Leyla's ten hours is exchanged for ten hours of Maria's physical labor. Under the circumstances, the only way to increase the total added value is to extend the hunting time of the physical labor. But by assumption, 10 hours' a day is the limit that can be employed and thus, the total output cannot be increased beyond its present level. The best the community can do is to re-produce the given value the following day. Their prosperity would never improve.

Assume that one day, one of the hunters; say Leyla, utilizing her mental faculties develops an "idea", a new hunting method (a new technology) which enables her to double the daily catch from 4 beavers to 8 beavers within the same 10 hour time-span. To be more specific, let us assume that she makes some simple tools to assist her in hunting the beavers. Due to the "idea", considered solely in terms of the physical labor time previously needed in trapping 4 beavers, the value of Leyla's new daily production of 8 beavers would increase from 10 to 20 hours of physical labor, although because of the new "production method" the hours actually employed are still 10.

New total supply/per day = 2 deer+8 beavers = 20 hours PL   (6)

But the total value generated is worth 30 hours PL

Or, alternatively

New total supply/day = 20 hours PL+ Leyla's ML   (7)

Leyla's ML has a daily of worth 10 hours of PL because her catch has doubled.

ML denotes "mental labor" or alternatively "productive knowledge", e.g., the new technology, while PL denotes physical labor. Leyla's mental contribution (new technology) is worth 10 hours of physical labor. In other words, additional-value is worth 10 hours of physical labor and in actual terms it means that Leyla's productivity has increased by 100 percent a day, in terms of PL. The community has become richer.

What would happen to the exchange relationship with the other hunter Maria now? Previously, there were 2 deer and 4 beavers in the market. Now, there are 2 deer and 8 beavers. With regard to the new situation, the exchange relations will have to change. What would the new exchange ratios look like?

"Unfair" Equilibrium

Case:1-A: Following in the footsteps of the Classical economists, one can argue that after Leyla's mental contribution, it still requires 10 hours' **PL** to catch 2 deer or, alternatively 8 beavers. Equal quantities of labor time are valid for both hunters, and therefore, 1 deer should exchange for 4 beavers instead of 2, in order to maintain the equality of the exchange when considering the labor-time employed. As a result, at the end of the day, Leyla would be expected to give up 4 beavers which are equal to 5 hours' physical labor in return for 1 deer, which equals Maria's 5 hours of physical labor a day.

This means that:

   Leyla's consumption should be 1 deer + 4 beavers = 10 hours' PL   (8)

   Maria's consumption should be 1 deer + 4 beavers = 10 hours' PL   (9)

But in terms of "initial values", each consumes products now worth 15 hours' PL instead of 10. Maria's 5 hours' physical labor a day could purchase 2 beavers, initially. Now, she can get 4 beavers; double the amount without any contribution on her part to the total wealth of the community.

Is this a "fair" and/or "rational" exchange?

If one ignores the "productivity increasing" contribution of Leyla's mental labor, exchanging 1 deer for 4 beavers would at first seem like an egalitarian exchange. But as yet Leyla has not been rewarded for her mental contribution which doubled her productivity and contribution to the common wealth. Instead of a combined value worth 20 hours' PL, there is now a total value which is worth 30 hours PL. Maria, the other hunter, who made no mental contribution to the common wealth would be the beneficiary of the new exchange relations based on the "physical labor-time consumed" approach. She works only 10 hours but consumes an output value worth 15 hours in terms of the physical labor-time employed. Meanwhile, Leyla who produces 20 hours value in terms physical labor-time only enjoys 15 hours' output. Such an exchange relationship would not encourage the further development of a new method of productivity, e.g., new technologies, since the system rewards the unproductive person, not the one who enhances the common wealth. In other words, the system is unable to provide any incentives for further value-generation and therefore the new exchange relationships are neither logical nor are they economically rational.

Unequal Exchange

Case:1-B: Initially, Maria and Leyla were exchanging 1 deer for 2 beavers. Assume that after the introduction of the new technology developed by Leyla, the initial exchange relationships are maintained. Maria and Leyla still exchange and consume 1 deer and 2 beavers each. But now, Leyla has access to an additional 4 beavers which are worth 10 hours' PL, which she can exchange for another product she wants, say for 2 sheep worth 10 hours' PL in another community. Maria still consumes 1 deer and 2 beavers (equivalent to 10 hours' PL) while Leyla now has 1 deer, 2 beavers and additional 2 sheep at her disposal for daily consumption. Leyla's mental contribution entitles her, given the demand, to a higher consumption level. Maria's standard is unchanged, but the community as a whole is more prosperous.

Leyla's consumption= 1deer + 2beavers + 2sheep = worth 20 hours' PL (10)
Maria's consumption = 1 deer + 2 beavers =   worth 10 hours' PL   (11)

In this case, there is no egalitarian exchange in the Classical tradition of equal quantities of labor expressed in terms of the time-unit employed. Nevertheless, neither Leyla nor Maria consumes less; in fact, there is an increase in total consumption due to Leyla's mental contribution. This outcome seems to be both, more logical and economically rational, than the foregoing Case:1-A.

A More Likely Outcome

Case:1-C: If there is an insufficient demand for Leyla's additional 4 beavers outside her own community, then even Maria might benefit from the new exchange relations and enjoy more consumption. Assume that the community consists only of Maria and herself and that Leyla can exchange only 2 beavers for 1 sheep outside her community. Leyla would now have 6 beavers at her disposal before entering the exchange relations with Maria, the other member of her community. If the market is to be cleared, Leyla will have to accept a new exchange relationship where 1 deer is exchanged for 3 beavers. Now, it is not only Leyla who is better off but so is Maria who actually made no mental contribution to the increased total supply.

Leyla's consumption = 1 deer + 3 beavers +1 sheep   (12)

Maria's consumption = 1 deer + 3 beavers   (13)

This outcome appear to be closer to reality than the two prior cases, because it allows even the less productive sectors of the economy to benefit from the overall development originating from other "productive" sectors. In other words, it is not only the inherently more dynamic industrial or manufacturing sector which benefit, the service sector, which is prone to a relatively lower productivity growth, can also benefit from a productivity growth in the other sectors.

Different Qualities of Mental Labor

In the simple model presented above, the increase in total wealth was the result of Leyla's "creative" contribution in the absence of any formal education or training. This creative feature of human mind helps to change and control our environment by creating "new" technologies. Only the human mind possesses the "creative" mental ability necessary to increase the productivity. Leyla's case was intended to demonstrate mental labor's contribution within the framework of a simple model. Her contribution and the contribution of countless others have been generated and accumulated for many centuries' even millenniums. This accumulated knowledge constitutes an immense pool presently at the service of mankind.

Nobody, no matter how brilliant his/her mental abilities are, acquires the knowledge in the form of "manna from heaven". In our age, knowledge is normally acquired through long years of formal or informal education/training and is enhanced by talent and experience which includes learning-by-doing. Personal abilities as well as the allocation of appropriate socio-economic conditions and opportunities naturally play a significant, if not a determining, role in the final quality of these abilities. Persons who are more fortunate than others in regard to their natural mental endowments or other man-made opportunities naturally acquire a higher degree of qualification than those less fortunate. The laborers are not a homogeneous entity; on the contrary, they are heterogeneous given the inherited natural abilities, they are also influenced by the socio-economic environment.

A contemporary labor force is expected to be able to make the best use of technologically sophisticated and complex production methods. But there is another and more significant contribution expected from the labor-power; that is the introduction of "new" and more advanced technologies. Therefore, it would be more appropriate to analyze the laborer's contribution under two different headings:

1- To maximize output with "given" technologies and resources (efficiency or micro-productivity analysis); and

2- To produce "new" products and/or production methods (technological productivity or macro-productivity analysis).

The impact of macro-productivity is the ever increasing wealth of nations and individuals, while the former, i.e., making the best use of "given" technology and resources, has only a limited impact.

Consequently, we can conclude that, given the natural endowments, the creative capacity of mental labor assisted by physical labor, is the only source of added value to all of the past and present value that has been generated so far and of any future growth. Or, to put it in William Petty's terms;

> "... labor is the father of material wealth, the earth is its mother." (in Marx, Vol.I, pp.133-134)

Keynes, unlike the Neoclassicals, had no problem with this notion.

> "I sympathise ... with the pre-classical doctrine that everything is produced by labour..... It is preferable to regard labour ... as the sole factor of production." (Keynes, 1991,pp.213-214)

Value-Price Relation

How are these values transformed into prices? That was one the central issues troubling the minds of the Classical economists. Ricardo had searched for an "invariable measure" of value to but could not find one, which satisfied him. He claimed that;

> "... there is no commodity which is not itself exposed to the same variations as the things, the value of which is to be ascertained; that is, there is none which is not subject to require more or less labor for its production." (Ricardo, 1990, pp.44-45)

For Marx, using the same exchange-value relations developed by Ricardo, the answer was obvious but Ricardo was unaware of his own discovery. The invariable measure Ricardo was looking for was the services of labor-power, which Marx defined as;

> "... the aggregate of those mental and physical capabilities existing in the physical form, the living personality." (Marx, Vol. I, p.270)

Marx had attempted to reduce the labor-power with all its physical and mental abilities into a simple quantitative concept in terms of the "socially necessary labor" measurable by the hours consumed and neglecting or overlooking the contribution of "creative" mental labor, which is a result of the labor power's mental abilities. As a result, the exchange relations as in the Case:1 above, where equal quantities of labor time consumed were exchanged, appears to be an egalitarian exchange relation.

As we have seen in the previous sections, given nature's indispensable role in production, the mental labor with its distinctive creative abilities is the only source of our being able to increase value. Therefore, the exchange relations in a proper "relative" value or price theory have to be based on a labor embodied approach with its dual properties, both the mental and physical aspects. Estimation of the actual amount of physical labor consumed is relatively easy to quantify. But, where to find a proper unit of measurement capable of accounting for the contribution of the mental abilities employed is a much more complicated task. Is it in fact a possible task? If there is no unit which we can use to transform these values into prices in a dynamic economy where the introduction of new technologies is a constant process, would demand schedule provide a resolution?

Relative Prices

Value of a product is the "value transmitted" to the product, given the demand for that product. In our simple model in Case:1-A, the relative values were determined by the physical-labor expended. But external demand as in Case:1-B and the external-internal demand relationships as in Case:1-C, showed that the magnitude of the demand is an important element in the determination of relative values. Bearing in mind that the accurate measurement of the value transmitted by the mental labor is highly unlikely, the relative market prices will be "assumed" to reflect the values transmitted and the magnitude of the demand. It is important to note that relative price ratios do not reflect actual transactions in a monetary economy accurately. But they can be used as a tool to demonstrate the crucial role of the mental contribution, in the form of "new" technology, in the formation of emerging new price levels, given the demand.

Case: 2-A (as presented above)

Let us start by reconsidering our simple economy with two hunters and introduce money as the sole medium of exchange in their transactions. Ignoring aspects like risk and profit for the sake of simplicity, and assume that one deer is worth 30 $ and one beaver 15 $. Initial exchange relations based on 10 hours' physical labor a day can be expressed as follows:

$$2 \text{ (deer)} * 30 \text{ \$} = 4 \text{ (beavers)} * 15 \text{ \$} \qquad (14)$$

Where;

1 deer = 2 beavers (15)

Or

30 $ = 2 * 15 $ (16)

Now, let us assume once again that Leyla, the beaver hunter, utilizing her mental faculties, develops a new hunting method which doubles her productivity from 4 to 8 beavers within the same time-span of 10-hours. Disregarding any reward for Leyla for her productive contribution, and estimating the value created by labor in terms of the time-units consumed, the new exchange relation between Maria and Leyla would be as follows:

2 (deer) * 30 $ = 8 (beavers) * 7.5 $ (17)

Where;

1 deer = 4 beavers (18)

Or;

30 $ = 4 * 7.5 $ (19)

In terms of initial prices, Maria's total labor valued at 30 $ now commands a 4 beaver value of 60 $. Is this a "fair", "logical" and/or "rational" exchange relation?

The equal labor-time approach to exchange rewards the less productive hunter, Maria, and penalizes the more productive Leyla. Under these circumstances, there would be no incentive for Leyla to make any effort to further improve her productivity. Naturally, a person might also be driven by motives other than financial reward. But for the sake of argument, we shall ignore such cases.

Case: 2-B

Given the initial price and demand where 1 deer is exchanged for 2 beavers, Leyla, the more productive hunter, could be in a better-off position if she can sell the additional 4 beavers in other markets. Given a demand by a third party consumer at the initial price of the beavers (15 $ each), Leyla's total income could increase from 60 $ to 120 $ a day, while that of Maria, the less productive one, remains at 60 $ a day.

Leyla's income= 1 deer (30 $) + 6 beavers (6 * 15 $)= 120 $ (20)

Maria's income= 1 deer (30 $) + 2 beavers (2 * 15 $)= 60 $ (21)

Meanwhile, the total income of both, Maria and Leyla, would increase from 120 $ to 180 $ thanks to the contribution of Leyla's productive mental abilities. Leyla's greater income is the result of and a justified reward for her intellectual contribution, e.g., productivity growth.

New total income/per day= 8 beavers*15$ + 2 deer*30$ = 180 $ (22)

The prices have remained unchanged i.e., at the initial level due to a sustained demand from third party sources outside the original community.

Case: 2-C

Suppose that the external demand is such that it causes the price of one beaver to decline from 15 $ to 10 $. As a result, and in order to clear the markets, one deer will have to be exchanged for 3 beavers in our original community. The new but somewhat deteriorated exchange ratio for Leyla would be as follows;

$$1 \text{ (deer)} * 30 \text{ \$} = 3 \text{ (beavers)} * 10 \text{ \$} \tag{23}$$

Plus two beavers worth 20 $ sold at external markets.

$$\text{Leyla's new income} = 6 * 10 \text{ \$} + 2 * 10\text{\$} = 80 \text{ \$} \tag{24}$$

$$\text{Total income} = \text{Leyla's income } 80 \text{ \$} + \text{Maria's income } 60 \text{ \$} = 140 \text{ \$} \tag{25}$$

Maria now consumes 3 beavers instead of 2, a 50 % improvement in her consumption. Maria and Leyla's joint total income is now 140 $. As a result, the deer hunter benefits from Leyla's mental contribution but Leyla end up with deterioration in terms of trade. On the other hand, both Leyla and Maria are now better off.

Relative Prices in the Service Sector

In the view of many Classical economists including Marx, the output of the service sector was considered as "unproductive". But, in the modern economies, it is an acknowledged and quantitatively proven fact that the service sector output is not only productive (value-producing) but also constitutes the largest share of GDP or GNP. There is nothing tangible or storable produced by the service sector, as is produced in the industrial sectors. Thus, there are no physical quantities to exchange as in the simple model above. In commodity production, there is a close relationship between the "tangible" input and "tangible" output, which normally move in the same direction. In other words, with given fixed costs, each unit output's cost is closely related to cost of the material inputs. In contrast, the service sector output is characterized by a higher intensity in the labor-power services. Unit costs are closely related to labor-power costs as a function of the time consumed, given the initial fixed cost and capital-goods cost. For instance, a teacher, a hairdresser, a business consultant or a musician can increase the total amount of the service they supply by working longer hours per day or week or month given the initial combination of the physical inputs which they employ.

By reducing the contribution of the labor inputs into a standard measurable unit such as the "labor-time-consumed" might appear to be an appropriate method with which to analyze the relative values or the prices used in the service sector. But, if we take into consideration the different qualities of mental labor required in producing the different kinds of the services that are required, such comparisons lose all their credibility. The value or price of the variable quality of the labor services will of course be different for each type of the service demanded. Thus, the value or price of each specific labor service would be different, even if the "equal labor time-consumed" unit was used in calculating the price or value of the supply for each individual service. Despite its significance for all national economies, a separate value/price analysis in the service sector will not be undertaken in this article, because of the limited scope of this article.

Commodity Sector Price Formation

So far, the analysis was focused on the creation and exchange of **relative-values or prices** in the commodity producing (tangible) sector. But, the analysis fails to reflect the real situation adequately; because the role of profit involved in the production process, in price formation and in exchange relationships have been neglected. In this part, profit will be introduced into the price formation analysis as it would occur in a monetary exchange economy. Introducing profits would inevitably lead to the simultaneous study of functional income distribution, which, however, will not be undertaken at this time due to limited space. The formation of new prices after the introduction of new technologies will be dealt with only briefly, again as a result of limited space.

Transformation of Values into Prices

A Case of "Barter-Exchange"

In barter exchange relations as in the simple model previously presented, the exchange-value of each product was determined by the mental and physical labor embodied in it i.e., (**LE**), the objective value (**OV**), the magnitude of demand for the product (**D**), and the subjective value (**SV**), which was previously assumed to be a "given". The last (**SV**) reflected the value that the end-users are willing to give up in exchange for a desired product, while the the objective value, **OV**, reflected the value of the past (**LE$_{t-1}$**) and the present (**LE$_t$**) mental and physical labor embodied in the product (**LE**), excluding any or all profits.

Initially, one deer was exchanged for two beavers. But after a mental contribution, Leyla's acquisitions had doubled and in order to clear the market, one deer had to be exchanged for four beavers, as in Case:1-A and Case:2-A.

In Case:1-B and Case:2-B, there was a demand for an additional four beavers which originated outside the community and as a result, Leyla's income doubled. From a more realistic and rational "exchange relation" perspective, which is found in Case: 1-C and Case: 2-C, both members of the community benefited from Leyla's mental contribution, but Leyla's profit was higher than Maria's. All these factors indicated that the market exchange-value (**MV**) of a product is determined by the **LE** and **D**. Thus, in the absence of profit, the **MV** equation can be shown as;

$$MV = f\,(OV\,;\,SV) \tag{26}$$

Or, alternatively

$$MV = f\,(LE\,;\,D) \tag{27}$$

implying that **LE** is the sole source of any value generation, while **D** gives the final shape to the exchange relationships by means of "haggling and bargaining". There were no profits involved and exchange was based on equal quantities of **MVs**, as in Case:1-C.

$$\mathbf{1\ MV^d = 3\ MV^b} \tag{28}$$

Instead of the initial

$$\mathbf{1\ MV^d = 2\ MV^b} \tag{29}$$

MV^d and **MV^b** denote the market values of deer and beaver, respectively. The critical question in these relative exchange relationships is:

How to transform these values into market prices (MPs) in the "barter" exchange?

To obtain the **MP** we simply have to add any profits (π) to both sides of the equation used in the exchange relationships. Since profit rates are expected to be equal as a result of competition, new exchange relationships would not affect the essential exchange and be based upon, in a sense, the original **MV**s of deer and beaver respectively.

$$\mathbf{MP^d = MV^d + \pi^d} \qquad \text{one deer's MP} \tag{30}$$
$$\mathbf{MP^b = MV^b + \pi^b} \qquad \text{one beaver's MP} \tag{31}$$
$$\mathbf{\pi = f\,(SV)} \qquad \text{or} \qquad \pi = f\,(D) \tag{32}$$

And, given Case:1-C;

$$1MV^d + \pi^d = 3\ MV^b + \pi^b \tag{33}$$

By assumption;

$$\pi^d = \pi^b \tag{34}$$

Then,

$$1MV^d = 3\ MV^b$$

As in (Eq.28), or, alternatively

1MP^d = 3 MP^b (35)

In other words, in a "barter" exchange model **MP**s deviate from **MV**s only to the extent of the size of the profits. Since profit rates are assumed to be uniform, **MP**s would reflect actual **MV**s. But, as we all know, barter exchange is a rare exception in modern economies.

Price Formation in a Monetary Economy

In monetary economies, given a fair competitive environment and an appropriate institutional and cultural setting, prices are determined by each firm in every sector on the initial costs plus a mark-up basis in accordance with long term profit maximization goals, given the level of demand. Costs include the prices of all the inputs of production including any profits along with the present wage level. There are no homogeneous products and thus no uniform price, even within the same sectors or sub-sectors. Accordingly, there are no homogeneous production methods, either. Each product can be produced by a different production method and displays unique product-specific features. In other words, each firm may require labor-power services as well as financial, organizational and technological settings in different qualities and quantities. Given these features, each firm would have different break-even points and different optimum plant capacities. As distinct from the simple model above, we have to bear in mind that there may be at least one additional input of production other than the input of the labor-power services.

Although labor-power is the sole source of all value generation, as before, the **MP** paid by end-users normally exceeds the value transmitted to the product. The reason for the difference is the **"profit"** (π) paid to the entrepreneur for the **"risk"** he/she assumed. This is not payment for a value transmitted to the product by capital. In fact, it is a payment "in excess" of the costs. In a sense, it is a **"surplus cost"** but not an **"unpaid surplus value"** to labor-power because the labor-power receives pre-determined wages in return of their services. This additional payment or **"surplus cost"** is a necessary and indispensable ingredient so that the capitalist system can keep functioning. Thus, the **MP** can be defined as **"the monetary expression of a product regulated by CP plus π (surplus cost) which is shaped by D"**.

Assume an economy where labor-power is the only input of production of a Commodity-**X**. Given **D**, the cost of production (**CP**) of **X** would be determined by;

$$CP^x = w * L \tag{36}$$

And;

$$MP^x = CP^x + \pi^x \tag{37}$$

In this initial stage of production with "labor-power input only", the **MV** of a product equals the **CP,** while the MP is greater than the MV **(MP > MV)** by virtue of the size of the "profit". In other words, whenever profits are introduced, neither **CP** nor **MV** would no longer reflect the true **MP** of the product.

Production with Multiple Inputs

Assume that Commodity-**X** produced yesterday is used as an input in the supply of Commodity-**Y** today. The **CP** of **Y** would comprise of past and present **w*L** plus the past π acquired;

$$CP^y = w^x * L^x + \pi^x + w^y * L^y \tag{38}$$

But,

$$MV^y = w^x * L^x + w^y * L^y \tag{39}$$

And, bearing in mind that **D** effects both π and the relative exchange ratios as in the simple model;

$$MP^y = w^x * L^x + \pi^x + w^y * L^y + \pi^y \tag{40}$$

To put it differently;

$$MP^y = CP^y + \pi^y \tag{41}$$

Where,

$$MP^y > MV^y \qquad \text{by the amount of} \quad \pi^x + \pi^y$$

Alternatively;

$$MP_t = wL_{t-1} + \pi_{t-1} + wL_t + \pi_t \tag{42}$$

t denotes the time.

In regard to a product with "**n**" inputs, the **MV** and **MP** would be:

$$MV^n = \Sigma LE^n \tag{43}$$

$$MP^n = \Sigma LE^n + \Sigma \pi^n \tag{44}$$

There is a break-even price level (**B**) where cost equals income. However, producers guided by long term profit motive would be reluctant to produce at price **B**, at least, not in the long term. Although **MP** always includes payment in excess of the **MV** of a product under normal conditions, the exchange of two producer-consumers, like Leyla and Maria, might present a different and interesting result. The producers would be exchanging equal quantities of **MV**s among themselves in a "**barter-like**" manner. In other words, since $\pi^a = \pi^b$, then **CP^a=CP^b** and accordingly, **MP^a=MP^b**. In this sense, they would be paying the break-even price, without making profits.

Meanwhile, however, the rest, or the non-capitalist end-users, have to pay profits in excess of **CP** for the system to function, which leads to different implications with regard to income distribution. It implies that some value above and beyond the production cost is being created by the non-capital-owners for the capital-owners.

Now, in the light of above arguments, the relevant and critical question is; where to set the **MP** in a competitive market for a given product with regard to the mental contribution, e.g., the technological innovations? Below, three different cases of price formation will be studied.

1-A Given product and a **given** technology (production method):

2-A Given product but a **new** production method introduced by mental labor:

3- New products supplied by **new** production methods.

A "Given" Product & a "given Production Method in regard to Price Formation

Let us first consider a case with a **given** technology embodied in the capital goods disregarding all non-profit motives. Given optimum plant capacity with a **CP** schedule and assuming optimum micro-productivity, i.e., optimum allocation of the inputs of production, the rate of profit (**r**) would be determined by **MP** set on a (**CP+π**) basis where π would be subject to the magnitude of demand (**D**). Or, to put it differently, the projected size or rate of profit by the firm would determine the **MP** and the quantity demanded or supplied (**q**), given the **PC** and the **CP**-schedule.

Figure-1 shows the relationship between the quantities supplied and profits with a given plant capacity. **B** denotes the break-even point of production where average unit costs equal the average unit returns, resulting in no profits. The size as well as the rate of the profit per unit output would increase as the total quantity supplied moves to the right of **q** towards **q***. Given **D**, at the maximum output level (**q***) determined by the plant capacity, the size and rate of profit would be at the highest possible level.

Assuming a uniform price for a product in a specific sector and "given" but "heterogeneous" production methods, **CP**-schedules and plant capacities. Profit rate for each firm might differ from the average profit rate in the sector as the quantities demanded or supplied vary from firm to firm, cet. par. In other words, similar products produced by different types of technologies would naturally produce different unit production costs, optimum plant sizes, break-even points and size or rate of profits. Differing levels of the quantities produced would also influence the size and rate of profits.

80

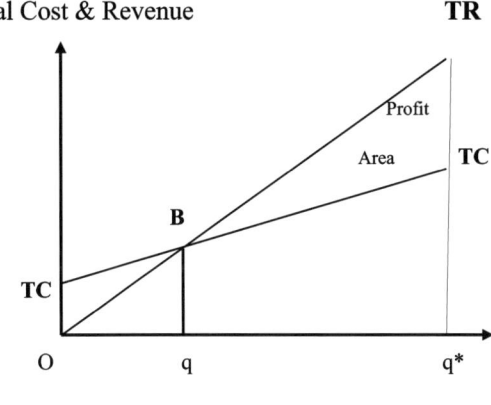

Figure: 1

How Influential is Variations in Demand?

Assume that the **MP** is initially set at a level, which assured a sufficient demand for full plant capacity utilization. But, then, for some reason the demand curve shifts towards the origin, from D_1D_1 to D_2D_2. Given the **MP** and full plant output capacity by **q***, as in Figure-2, the shift in the demand curve would cause alterations not only in the quantities produced but also in the size and rate of the profits. The profit would follow a decline in **D** and drop from the area of the quadrilateral "**abcd**", to the area of "**abB**", also causing a decline in the quantity produced by the gap of q^1q^x, e.g., $Oq^x - Oq^1$. The new but inefficient capacity utilization level is denoted by the dashed line aq^1. At that level, both the size as well as the rate of profit would be lower.

Following the decline in demand and the ensuing excess plant capacity, the firm may respond by changing its price. If the **MP** is increased, the revenue line **OTR** would become steeper, OTR^1, but it would be unlikely to restore the initial profit size and full plant capacity level (Figure:3). If the **MP** is reduced in response to a fall in the demand curve, the **OTR** revenue line would become flatter; OTR^2, and the plant capacity utilization would increase, if the end-users respond by increasing their demand. But the restoration of the initial profit level would be impossible.

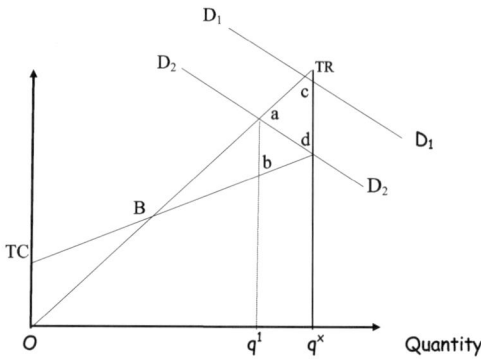

Figure: 2 Total Cost & Revenue

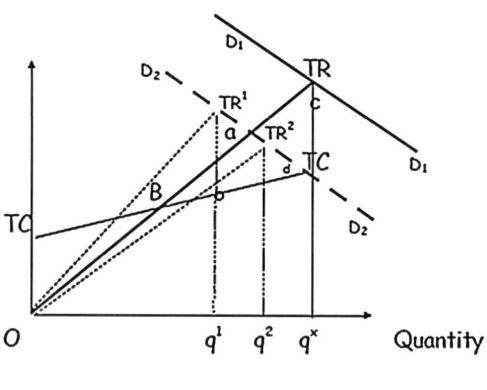

Figure: 3

To conclude; the labor-embodied (**LE**) argument alone falls short of explaining the market prices, especially in the case of a supply-demand imbalance. **CP**, which reflects (past and present) **LE** and past π seems to regulate the minimum **MP**- level, which, in the final stage, is adjusted by the "haggling and bargaining" in the market. Thus, fluctuations in demand causing imbalance in supply-demand conditions have an influential impact on the determination of the short term **MP**. With a "given" technology and a supply-demand balance, the long term **MP** is more likely to reflect the **LE** and pastπ, plus the present π shaped by present **D**-schedule. But, there is a crucial reality; technology (mental contribution) is not a static "given", on the contrary, it is continually changing.

Technological- (Macro-) Productivity Growth & Price

The major distinction of the following analysis is the introduction of "new" technologies, which are the products of the creative mental labor. By assumption, in this analysis the demand is a given" and there is no excess plant capacity.

A "Given" product but "new" production method and price

There are two motives for a profit driven firm to produce a "given" product with a "new" technology; either (a) to make higher profits by reducing unit **CP**, and/or (b) to become more price-competitive. After the introduction of new method; the **expected** and normally **realized** rate of profit should be higher than before, at least until the others catch up. Figure:4 shows a hypothetical case of declining costs and increasing profits in relation to a cost-saving technological change with a given output. TC^x-TC^x line indicates the new cost curve, which is now closer to the origin as a result of the new method indicating lower production costs, thus higher profits. The new profit area is **acB^x** where **acB^x>abB.** The profit area before the introduction of the new technology (productive knowledge) is indicated by the triangle **abB.**

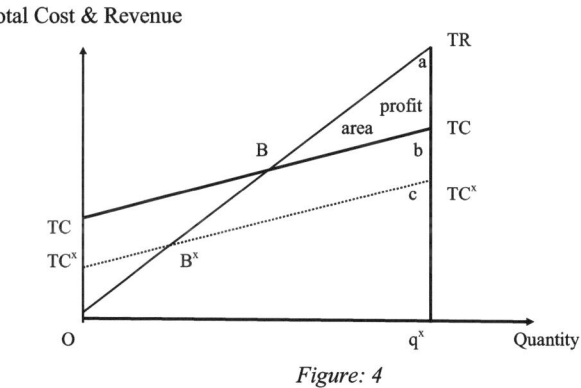

Figure: 4

If the "new" technology facilitates an increased supply with "given" inputs, then, again, the size and the rate of profits would increase, cet. par. (Figure:5). The new plant capacity is denoted by **Oq^{xx}** while **BTR^{xx} TC^{xx}** denotes the new profit area, which is larger than **BTRTC.** The new situation also implies a lower **CP** per unit output as the share of fixed costs in the per unit output declines.

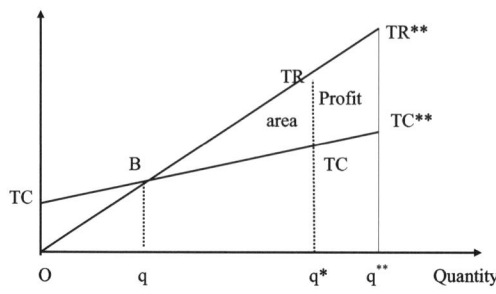

Figure: 5

Now that the unit production costs are lower, will the price remain unchanged?

If there is a sustained demand at the initial price level, there would be no need to change the price, as shown in the Cases 1-B and 2-B in the simple relative exchange model with our two hunters. As a result of new technology, the size and the rate of profits would increase. If the firm wants to improve its competitive position, then it would have to lower the price. The long term targets of firms would determine the limit of price decline. Reducing the price until the previous level of profit rate is restored would not be an irrational behavior, cet. par.

Conclusion; as a result of the creative mental labor's contribution, the new price level will have to be set somewhere between the prevailing **MP** and a lower level above break-even price in accordance with firm's short and long term targets and the magnitude of demand.

"New" products - production methods and price

The distinctive feature of economies is the continuous introduction of "new" products as well as new production methods, which appears to have accelerated in the present so called "the age of knowledge". Since both, the products as well as the production methods are **new**; there would be no preceding price level with which to make a comparison. In other words, a study of the effect of the price of a **new technology** on a **new product or production method** would only give us information on the "new" sets of prices. It can be claimed, however, without any hesitation, that the **expected**, and normally the **realized**, profit rate would be higher than the prevailing average market rate, because the owner of the new technology would enjoy a monopolistic power in the market, until the competitors catch up.

Concluding Remarks

The purpose of this paper was to display the genesis of and the ever changing source of "added value". The analysis so far has indicated that, given natural resources and physical labor, **the creative knowledge of labor-power** is **the sole source of all the exchange-values created and all the wealth which is accu**mulated. Price is determined by three factors; cost of production, profit rate and demand.

In a capitalist market economy where money is used as the medium of exchange, the capital-owners obtain a part of the sale-price referred to as "profit". What capital-owners receive as profit is not a part of the income generated by capital-owners. It is a part of the added value paid to capital-owners (investors) as a reward the risks they have assumed in the supply of these goods and services, which is both legitimate and necessary for the smooth functioning of markets to enable them to supply the products which are in demand.

The rate of profit always affects the sale price directly and influences the trend of demand. On the other hand, the degree of demand may in turn influence the sale-price of products as well. To put it another way, the sale-price influences the level of demand while the level of demand may influence the sale-price; there is a reciprocal interaction.

All commodities are originally the product of nature transformed or reshaped by labor-power. Given the limited impact of the physical labor, the creative mental faculty of labor-power continually introduces new ideas in order to change and control our environment, which in its turn changes our entire way of life. Assuming an optimum resource (micro-economic) efficiency in production, mental labor's "new" contribution that is "new" technology influences the price-level by introducing either;

    1- a cost-saving "new" method of production, "given" the product; or

    2- "new" products with "new" sets of values and prices.

In the first case, the potential profit rate, and in the second, the expected, and the usually realized, profit rate is higher than the average rate.

Unequal distribution of income has always been a major problem area and an embarrassment for both economic science and economists. The analysis above indicates that this problem can be tackled, at least to some extent, by increasing the number of persons in a "barter-like" exchange, that is by making people "profit receivers" who exchange products with profits instead of being just wage-earners who have to pay in excess of the value transmitted to products. In a society where everybody receives an "equal" share of profits, income distribution would be much fairer, though not at an optimum level.

References

Blaug, M.   1990   *The History of Economic Thought*
                   Edward Elgar Pub. Ltd., Hants
Gürak, H.   2000   "Economic Growth and Productive Knowledge"
                   YK-Economic Review, Vol. 11, No:1, pp. 55-69.
Hicks, J.   1983   *Classics and Moderns*
                   Basil Blackwell Publ., Oxford
Keynes, J.M.1973   The General Theory Of Employment, Interest And
                   Money, Macmillan Cambridge University Press
Marx, K.   1976   *Capital*, Vol. I
                   Penguin Books
Marshall, A. 1990 *Principles Of Economics*, Vol. I & II,
                   Macmillan And Co., London
Ricardo, D. 1990 On The Principles Of Political Economy And Taxation,
                   Cambridge Uni. Press
Schumpeter, J.A. 1954   *History of Economic Analysis*
                         Oxford Uni. Press, New York

3- THE CREATIVE MIND & NEW TECHNOLOGIES

The genesis & the engine of the prosperity of nations

> "Knowledge is the most powerful engine of production; it enables us to subdue Nature and satisfy our wants."
> Alfred Marshall

Introduction

How an economy grows has always been a fascinating subject for the economists. For decades and for centuries numerous attempts have been made to identify the factors determining growth by many able minded scholars like A. Smith, Marx, Solow, et al to find universally applicable explanations which can be applied to the subject. Smith's analysis on the division of labor and the correlated productivity increase, Ricardo's remarks that the primitive hunter's capital good (the weapon) was a product of the laborer and Marx' comments that capitalism's internal forces in turn leads to constant technological change were all steps in this direction. But eventually, the determining factors of growth were reduced to two major factors by the neoclassical school i.e., "capital accumulation" and "population growth" which were used for decades to follow.

For a long time, the textbooks on economic growth influenced by the neoclassical doctrine were used to inform others that this growth process was influenced by two factors;

1. Endogenous factor (investments),.
2. Exogenous factor (population growth).

The endogenous factor was assumed to bring the economy eventually to a stationary equilibrium as the neoclassical growth models predicted. Once in equilibrium, the only factor causing growth would be the exogenous ones. The model was quite logical and consistent but at the same time, unfortunately, short on realism because, amongst other things, it failed to embody "technological progress" as an endogenous factor of growth.

Contemporary theories of growth seem to have focused on two central questions:

1. Why do economies grow in the long term?
2. Why do economies grow at different rates?

The first requires the objective and scientific study of the nature of growth, e.g. globally valid sources and to determine the essential factors of growth, while the latter emphasizes the comparative analysis of different growth rates and their determining factors, in regard to historical developments, the institutional and cultural infrastructure and present economic policies..

Which one is the right question to start with? The immediate and intuitive reply would be that both are important in the study of the growth process, though the former seems to deserve more credit. But before attempting to answer these questions, we need to know about the initial source of the long term growth process, i.e. **the genesis of growth**. In other words, we have to have some knowledge about the initial stages of growth, before going into a long term growth analysis.

Do we in fact have a logical and consistent theory of growth with the appropriate explanatory power about the genesis of growth, which is indeed capable of accounting for both the causes and the consequences of real world transactions? Unfortunately, the answer to this question is that we do not.

The Subject of this Study

The central theme of this work is about the **Genesis of Growth**? While making an attempt to analyze and explain the nature of the initial determinants of growth, it will be inevitable to take up the issue of the **Genesis of Value Generation**, as both take place simultaneously. In other words, this paper shall make an attempt to present a coherent, consistent, logical and, at the same time, a realistic account of **value generation and long term growth**.

The central hypothesis is that all economic progress (or growth or improved living standards including wealth) is based on technological change as Marx, Schumpeter and many others would agree. But the more accurate and crucial assertion would be that the origin of all technological change is "productive knowledge" or synonymously "knowledge of production" which is the **product of the creative human mind**", i.e., the "mental labor".

Thus, technological change originating from the creative human mind (mental labor) has always been the main source of all economic growth and value generation in both the past and present and will continue to be so in the future. In other words, the genesis and the generator of all-new products and new production methods is the creative knowledge of human beings, given, of course, "the benefits of nature", the existing level of knowledge and the institutional and cultural settings.

The Study Plan

Before attempting to present a simple model of value generation and the genesis of growth, some key concepts like knowledge, science and technology will be discussed. Then, a brief review of economic ideas on growth and technology will follow. And finally, a simple model based on productive knowledge, i.e., technology, will be presented that will account for the genesis of value generation and growth.

Knowledge - Science & Technology

Ideas can only be understood correctly if the correct definitions and expressions are used to explain them. Therefore, there seems to be a need for us to define key concepts such as "knowledge", "science" and "technology" to avoid any misunderstanding arising from divergent interpretations.

Knowledge

According to P. Drucker and many others, the most important input of production is "**knowledge**". But what "kind" of knowledge is meant? Does it imply "general knowledge" about history, geography, literature, etc., or some "specific" knowledge related to production, i.e., **technology**? Certainly Drucker appears to mean the latter though it is not stated explicitly. A more explicit account of what he meant would avoid confusion and the correlation between this specific kind of knowledge and the growth of productivity (which is highly praised by A. Toffler, P. Drucker, R. Lucas, P. Romer, et al) could be better evaluated.

Knowledge can be defined as a set of "expressions" based on facts and understanding acquired through education, observation, experiment and experience. In other words, knowledge is an intellectual product of the creative human brain which continuously adapts to changing conditions and experiences.

There is a distinction between "information" and "knowledge". Often people use the term "information" as if it were synonymous with "knowledge". **Information** seeks answers to questions of a "descriptive" nature. For instance: "What color is the sky"? "Where does the President live?", or "What is on TV tonight?" It is normally acquired from the media or an acquaintance or from some form of awareness or association. Information as such does not directly contribute to either the explanation of facts or of incidents or to the increase in prosperity of individuals or nations but might indirectly influence the receptive and the creative capacity of the human brain.

Knowledge can be divided into three categories;
1. General knowledge.
2. Scientific knowledge.
3. Productive knowledge (knowledge internalized in products and/or methods of production; i.e., **technology**).

General knowledge is beyond the scope of our interest here.

Science or Scientific Knowledge

Science can be defined as **"the pool of explanatory knowledge"** which can be used to better understand and control the environment we live in, and often provides commercial interests with the essential ingredients to invent and develop new products and production processes. It aims to discover the unknown truth behind facts and phenomena such as whether there is life in other planets or the evolution of certain species.

There are some similarities but no clear distinctions between the concept of science and technology, except for the distinct profit motive of the latter. In general, science seems to deal mainly with the acquisition of knowledge "per se" while technology is oriented towards the reshaping and controlling of the environment in order to improve living standards. The contribution of scientific research to technological knowledge and, in some cases vice-versa, is indisputable. **"Science without the by-play of technology becomes sterile while technology without science becomes moribund."** (Jones,1971,p.6).

Productive Knowledge (Technology)

Technology will be redefined as **"productive knowledge"**, that is as the set of specific knowledge used to supply products or the methods of production responding to the practical needs and wants of the community. A more concise definition would be: Technology is **"knowledge applied to resources through human labor and internalized in the products"** ranging from consumable products to machinery, semi-finished goods, tools, etc. As an UNCTAD study postulates:

> "Technology is the key to the progress of mankind and that all peoples have the right
> to benefit from the advances and developments in science and technology."
> (UNCTAD, 1983,p.1).

Technology does not only assist us in satisfying our needs and desires but it also increases our control over the environment in which we live. By combining the abilities of the labor services and natural inputs, technology improves the standard of living of human beings both qualitatively and quantitatively.

Technology is usually produced by people trained in disciplines that are oriented to the commercial exploitation of knowledge and sponsored by commercial interests. The greatest difference between science and technology can be described as follows; science aims to **"discover"** the facts of nature or the universe while technologies are **"created"** by mental labor in pursuit of profits for the producers. Technological findings are most likely to be patented by either the researcher or the sponsor or both as both expect returns i.e. rewards from the application of these technological findings. However, technology is not always fully utilized for the benefit of mankind. Especially in developing countries various factors are responsible for the limited utilization of the available technologies (Gürak;1990).

Technological changes are normally characterized by three phases all of each can be costly as well as resource and time consuming. These are:

1- The Research Stage,
2- The Invention Phase, and
3- The Innovation or Development Period.

The innovation or development of an invention is inherently a risky process. The innovation process is often a more time and resource consuming process than the initial research and invention phases. The longer it takes to develop, the higher the costs and risks will be. The technical and commercial circumstances are the major determinants of the developmental process. The data on past technological expenditure indicate that all major breakthroughs have consumed enormous human and financial resources as well as time. Some important innovations appear to follow an invention after a considerable period of time possibly even after several years have passed. This time lag seems to be shorter for consumer goods than for other methods of production.

Technology being the **"key to the progress of mankind"** is expected to pave the way for economic development and narrow the gap between the developed and developing countries. Not only in our present era, but also throughout the entire history of mankind productive knowledge has been the key to growth by increasing the productivity per employee. In other words, technology has always been the true source of improved standards of living even in pre-commercial societies where bartering was the common means of exchange. Technology operated the same way even in "hunter-gatherer" communities.

But there is something more important than the technology itself; that is its "source"; **the creative intellectual labor of human beings.** Human beings not only produce "productive knowledge (technology)", but also apply it to specific areas in order to supply various commodities and services. Let's assume that there is a grand pool of productive knowledge but no appropriate human resources to make use of it. Who would produce all the tangible as well as the intangible output? And more importantly, who would continue to replenish this pool of existing knowledge? Therefore, it would be more appropriate to say that, given the physical labor inputs, all the institutional and cultural infrastructure and the natural resources, the most important input of production throughout history has been the able minded human resources, i.e., **knowledgeable and creative human resources**, or in the neoclassical jargon, **"Human Capital"**.

By the way, why is it called "Human Capital"? Why not **"Intellectual Labor"** or **"Mental Labor"**? Could it be because of some clash of ideologies or, a cultural heritage that defines human beings who labor as "work horses" and thus consequently insignificant?

To summarize, the two most distinguishing features of labor force are;

1- The ability to create new technologies or productive knowledge; and

2- The ability to exploit existing technologies efficiently.

A Brief Historical Review

Nowadays, a growing number of economists acknowledge the contribution of productive knowledge and pay more attention to it. But what was the role of technology or mental labor in past economic analyses?

Classical economists were aware of the vital importance of both, mental labor and technological change. But, they were mainly interested in issues like value, exchange, wages, prices, profit rate, interest rate, accumulation and foreign trade. They, in general, had assumed that by accumulating more of the means of production, i.e., homogeneous capital goods, a nation would become richer. **The keyword of growth was "accumulation"**. Highly praised free international trade was assumed to expand the international markets as well as the domestic one, thus facilitating further investment and growth. But, by now we know, a nation cannot grow rich simply by accumulating more and more of the same physical capital goods.

Now, let us see what some prominent economists had to say about economic growth.

A. Smith

Smith (1976) had implanted the seeds of productivity growth into the economic theory by pointing out the importance of the division of labor in the creation of the wealth of nations:

> "The greatest improvement in the productivity powers of labor and the greater part of the skill, dexterity and judgment with which it is any where directed, or applied, seem to have been", he said, "the effects of the division of labor." (Smith, 1976,Vol.I,p.13)

But according to this definition, productivity increase was a result of the division of labor only. Technological change could, at best, follow the division of labor in order to adjust the production process to the new conditions. Surprisingly, to Smith, the division of labor was; **"not originally the effect of any human wisdom."** (Smith; 1976; Vol.I; p.25) It was the entrepreneurs' drive for profits that increased productivity through the division of labor that gave rise to technological change.

Ricardo

Ricardo's studies emphasized, like his contemporaries, issues such as value-generation, exchange, rent, profit rate, wage rate, accumulation and foreign trade. In the analysis of value generation and exchange, he was directly referring to the correlation between the hunter's (technology embodying) capital goods and increased value generation (growth). First of all, the "capital good" of the hunter is a product of mental labor, embodying knowledge in a physical shape. And, secondly, if the value generated is an addition to the available one, it implies growth. In Ricardo's words:

> "Without some weapon, neither the beaver nor the unit of Y could be destroyed, and therefore the value of these animals would be regulated, not solely by the time and labor necessary to their destruction, but also by the time and labor necessary for providing the hunter's capital, the weapon, by the aid of which their destruction was effected." (Ricardo,1990,p.23)

In fact, hunter's capital good, the weapon, was a product of the hunter's intellectual labor, which simply combined his intellectual and physical labor to rearrange nature's endowments. It was the supply of his capital good that eventually resulted in the increase in his productivity. But, Ricardo appears to have failed to acknowledge the impact of mental labor and technological innovations or, probably, overlooked these features and concentrated on other issues more "important" for him at the time.

According to Ricardo, there were increasing returns in the dynamic industrial sector due to constant technological innovation, but decreasing returns in the "static" agricultural sector. Overall, he assumed that, decreasing returns would be valid for the entire economy and the growth process would come to an end due to falling profit rates.

Marx

Marxist School, on the other hand, was preoccupied with the demonstration and proof of the doctrine's major claims such as the creation of surplus value, exploitation, the decreasing rate of profits and imperialism. Marx had pointed out an important and inherent feature of capitalism, the **"creative destruction"** of the system. The capitalists in fierce competition with each other were under constant pressure to find and introduce new technologies. This was the system's "progressive" but at the same time "destructive" feature. Unfortunately Marx referred to technological change just to display how the exploitation rate (surplus value) is increased. Marx and the Marxists seem to have overlooked or missed or played down this "creative destruction" which is an inherent feature of the capitalist system. If due attention were paid to the incessant technological changes, they probably would have discovered the essential role of technological change a long time ago and found out why the profit rates were not declining.

Marshall

Marshall was one of the prominent scholars to point out specifically the importance of "knowledge" in economic relationships. Decades ago he stated that:

"Knowledge is our most powerful engine of production." (Marshall,1990,p.115) "In a sense", he continued, "… there are only two agents of production, nature and man. Capital and organization are the result of the work of man aided by nature."
(Marshall, 1990,p.116)

But, unfortunately, he did not attempt to incorporate "knowledge" into a growth theory, either.

Keynes

For Keynes, economic growth was a matter of effective demand and investments (accumulation). The emphasis of the analysis was on unemployment equilibrium and its restoration to the full employment equilibrium. As output grew, it was assumed, further divisions of labor would increase productivity. Due to the assumed correlation between effective demand and growth, the role of technological change could not get the proper attention it deserved.

Schumpeter

Decades ago Schumpeter, in a fashion similar to some Classical economists, especially to Marx, pointed out that:

> "Capitalism ... is by nature a form or method of economic change and not only never is but never can be stationary." (1970,p.82)

And re-emphasized that the capitalist system;

> "... incessantly revolutionizes the economic structure FROM WITHIN, incessantly destroying the old one, incessantly creating a new one. This process of creative destruction is the essential fact about capitalism." (1970,p.83)

The result is a system of continued "disequilibrium" as change succeeds change in an uninterrupted fashion. This incessant revolution from within came, of course, through technological changes that implied new consumer goods as well as new methods of production. But Schumpeter, like Marshall, did made no attempt to incorporate technological change, the capitalist engine in motion, into a growth model.

Changing Trends in the 1950s

Ignorance of the impact of technology began to change with the contributions of scholars like Abramowitz, Denison, Solow, Schultz, Becker and others who emphasized technological change and intellectual labor, "human capital" in the Neoclassical jargon.

In a study on the growth of the USA, Solow found that basically growth was caused by technological change, rather than by increased capital and/or labor. His conclusions attracted more and more research on the role of technological change on growth and consolidated its place in economic analyses. Yet, technological change remained, for some time to follow, as an **exogenous** factor of growth, coming like manna from heaven. As a result, the equilibrium model was not disturbed but the economic analysis was not enriched, either.

Paul Romer

This unrealistic approach to the role of technology was, finally, abandoned by another prominent scholar of the neoclassical heritage, Paul Romer, whose contribution (1990) was characterized by the theory of **endogenous technological change**, as "**the engine of growth**". To reflect its divergence from the mainstream approaches was given the title of "**Endogenous Growth Theory**". His model emphasized the importance of "knowledge" by making knowledge a pivotal concept in his analysis. His endogenous growth approach has brought refreshingly new dimensions to discussions on growth theory. But it also contained some serious omissions. One of these was that though new technology meant a new idea or a new design, produced by human capital, he overlooked the fact that the concept of "human capital" refers to the "qualities" of laborer (L). In other words, human capital and laborer are not two distinct factors of production and that the knowledge produced by **L** is embodied in the capital-goods and the consumer-goods produced.

Increased Productivity (Growth)

Having elaborated on the concept "knowledge" and "knowledge of production", it will be appropriate to proceed with a related and important concept for our analysis, that of "increased productivity". Increased productivity has always been the source of increased economic welfare though the word in its present context could not even be found in economics lexicons until five or six decades ago. According to a widely accepted description, the concept of productivity refers to the relationship between output and the inputs of production. This is an unsatisfactory and misleading description of productivity. It would be a more realistic and appropriate approach to treat productivity as the "**ability to add value (wages and profit)**". "Productivity growth", on the other hand, refers to a dynamic process, and is synonymous with growth.

In the long term, productivity growth is always due to **new technologies** and is also referred to as "**technological-productivity**" or "**macro-productivity**". Given the technology, productivity can also be increased in the short or medium term by various other measures like improved education and training, reallocation of resources, reorganization of production, etc. This will be referred to as "**micro productivity**". Thus, the precise distinction between macro and micro productivity is that the former (macro-productivity) requires "new" technologies while the latter (micro-productivity) is realized with "given" technologies. According to this classification, "**resource efficiency**" which means the degree of utilization of the given human-physical and natural resources (rows 4-10 on Table-3) and **"economic efficiency"** are treated as **"micro productivity changes"** (see Table:1). In the absence of new technologies, micro productivity growth would in time reach its limits and the economy would enter a "stationary" stage with no growth at all. Therefore, macro productivity growth is a critical and important concept for long term prosperity. Every incremental increase in productivity implies increased value-generation (added value) with regard to the inputs thus ever increasing incomes and profits.

Table:1 Micro and Macro Productivity and Profitability

			Technology
Macro (technological) productivity	Monetary or Quantitative	TR/TC Q^s / Q^i *	New
Micro productivity (Economic efficiency)	Monetary	TR/TC	Given
Profit rate**	Monetary	π / K	Given or Variable

* Q^s: quantity supplied ; Q^i quantity of inputs; TR: total revenue; TC: total cost

** Profit rate = Profit / Capital Employed. Profits are maximized at max economic and resource efficiency.

Thanks to the continual macro productivity growth, mankind has been able to consume ever more and higher quality products ranging from manufactured goods to services and enjoy decreasing production costs per unit of output. In the absence of macro productivity growth, the magnitude of consumption would still be at the primitive subsistence or survival level, as it is in the animal world. Nowadays the concept productivity growth has a wide range of applications ranging from **tangible goods** e.g., mining, manufacturing and agriculture to **intangible service sector activities** such as health care, tourism, banking, etc.

With the introduction of a new technology, the producer expects to realize a higher profit rate than before. If the new technology is to produce a given product with a new production method, then unit costs are likely to decrease and raise the "rate of profit" (**r**). If the new technology introduces new products, then the "expected" profit rate is higher than the average rate for that particular sector. Otherwise, there would be no incentive to develop and introduce new technologies (see Table:2).

From the point of view of profit driven firms, the size or the rate of **VA** to costs is not as important as the rate (**r**) and the size of profits (**π**). But for the nation as a whole the aggregate added value (**GDP**) has more significance because it consists of total wages (**w**), profits (**π**) and interest (**i**). The higher the **GDP** the greater the per capita income of the nation will be and thus the purchasing power per capita will increase, cet. par.

When total added value is measured in regard to all inputs, it is called "Total Factor Productivity" (**TFP**) (**Added value/Labor inputs + other inputs**), which is not the same as Solow's **TFP**. If the productivity ratio is acquired against the use of one or more inputs, it is referred to as "Partial Factor Productivity" (**PFP**). The most common denominators used in **PFP** analysis are the labor input, i.e., labor productivity per hour or day (**LP**) (Value-added/Labor inputs) and the "Labor Wage Productivity" (**LWP**) (Value-added/Labor inputs*Wage rate).

Table: 2 Impacts of a New Production Method

Type of technological change	K-saving per unit q	Impact on Output	VA	Q/L	LWP	r	π/VA	W/VA
Labor saving	Yes	$q_{t+1} = q_t$	↑	↑	↑	↑	↑	↓
X_i saving (exc. L)	Yes	$q_{t+1} = q_t$	↑	The same	↑	↑	↑	↓
Q increasing	Yes	$q_{t+1} > q_t$	↑	↑	↑	↑	↑	↓
L & X_i saving while Q increasing.	Yes	$q_{t+1} > q_t$	↑	↑	↑	↑	↑	↓

- Prices (p), wage rate (w) and interest rate (i) constant
- Wage Bill (W=w*L)
- X_i = Inputs (excl. L)   i = 1,2,....,n
- TFP = VA / K   (incl. L)
- PFP = VA / X_i * p_i (excl. L)
- K = p_i X_i + wL (Capital advanced)
- VA = w*L + π (incl. Interest rate cost)
- r = π / K

The Quantity vs. Value Approach

There are two ways to measure productivity and productivity growth. The quantitative approach displays disadvantages with regard to the estimation of the productivity growth. For instance, a given amount of output may require less units of energy (**E**) compared to a previous production period ($E_{t+1} < E_t$). Though the quantity supplied has not changed ($Q_{t+1} = Q_t$) there would be no problem in estimating the productivity increase in regard to energy input ($Q_{t+1}/E_{t+1} > Q_t/E_t$). The same would apply if the energy input was substituted by a "labor unit" input. But how are we able to measure the quantitative productivity change in regard to two or more variable inputs, say, energy and labor?

Technological productivity analysis, on the other hand, takes into consideration "new" products and production methods. Since they are new, there is nothing with which to compare them or a set standard against which their technological productivity can be measured. The new technology will only lead to increases in the variety of goods and services supplied within a country. When Drucker (1995) pointed out that an annual rise in productivity of 3-4 percent was the basis of all improvements in the standards of living; he actually was referring to "macro" economic changes. As a result of continuous and successive productivity increases, i.e., technological changes, there are now ever more commodities and services, supplied with much less labor and/or other inputs than a century or a decade or even a year ago. "**Output per hour worked in the US today is 10 times as valuable as output per hour worked 100 years ago**." (Romer;1990).

How to Increase Productivity?

Productivity can be increased in many ways and 10 of them are shown on Table: 3. Rows **1** and **2** require **new technologies** to increase productivity. Row 3 indicates a unique situation; the transferred technology is a "known" one for the technology owner or seller, but implies a "new" technology for the buyer. The rest from 4 to 10 indicate that a rise in productivity can be achieved even without introducing new technologies. In all 10 cases the rate of profit is expected to rise as a result of productivity growth, cet. par. In Row-2, normally there would be no previous data with which to compare. All the variables would be expected to move in the directions of the arrows. However firms sometimes may introduce new technologies for defensive purposes in order to survive in the face of strong competition.

The Aggregate Added Value or synonymously the Gross Domestic Product at Factor Prices (**GDP**) indicates the sum of all the value added within a country. It is a useful criterion for a comparative analysis of the aggregate productivity, (**GDP/Per Employee**), which is also referred to as the "**National Economic Productivity (NEP)**" or alternatively, of the aggregate welfare (**GDP/Per capita**) which is referred to as the "**National Economic Welfare (NEW)**". Yet, these terms ought to be applied carefully. For example, as far as **NEP** is concerned, an estimation of productivity not infrequently refers to the sectors with only **tangible** outputs such as mining, manufacturing and agriculture. But **GDP** normally includes all kind activities, including the **intangible** service sector activities.

Comparison of per capita **GDP** is a frequently applied criterion in comparative analyses. But it has its shortcomings as well. For instance, domestic prices in Developing Countries do not always reflect their actual values in terms of purchasing power. Therefore, a purchasing power parity criterion might offer a better premise for comparative studies.

Table: 3 10 Different Ways to Increase the Productivity [a]

	New technology	Type of change	VA/K	VA/L	r	π/VA	W[b]/VA
1	Yes	**New Method of Production but given product**	↑	↑	↑	↑	↓
2	Yes	New Product & Prod. Method	↑*	↑*	↑*	↑*	↓*
3	No	Transfer of Technology	↑	↑	↑	↑	↓
4	No	Reorganization	↑	↑	↑	↑	↓
5	No	Increased Capacity Util.	↑	↑	↑	↑	↓
6	No	Shift-work	↑	↑	↑	↑	↓
7	No	Reallocation	↑	↑	↑	↑	↓
8	No	Gen. Education & Training	↑	↑	↑	↑	↓
9	No	On-the-job Training	↑	↑	↑	↑	↓
10	No	Improved Health-Safety	↑	↑	↑	↑	↓

a Prices (p), wage rate (w) and interest rate (i) constant

b Wage Bill (W=w*L)

* There is no previous data for a comparison. But the expectations of the entrepreneurs would be in the direction of the arrows.

The Genesis of Productivity Growth

If productivity growth, whether it be at a micro or a macro level, is the source of a nation's increasing standard of living then **what is the genesis of this productivity growth?** Since productivity growth does not come like manna from heaven, there has to be an economically rational explanation concerning its origin.

New knowledge or new technologies are always **"the products of the human mind**. The productivity of the mind is directly associated with and the application of the available pool of accumulated human knowledge, education-training, individual talents, experience, the scientific-technological infrastructure, the institutional setting and the prevailing economic policies.

Do We Need a Value-Price Theory Embodying Productive Knowledge?

If one acknowledges that advances in productive knowledge (technology) are the origin of all the added value and accumulated wealth, given nature's endowments, a related and extremely critical question rears its ugly head. Namely: **what is the role of productive knowledge in the value-price theory?**

After criticizing the economic theories for their inadequacy in explaining actual economies, Drucker states that none of the great non-Marxist economists of the past 100 years e.g., Marshall, Schumpeter and Keynes were;

> "… comfortable with an economics that lacked a Theory of Value altogether. But as Keynes anecdote illustrates, they saw no alternative." (1981,p.21)

Economic science still lacks an **"acknowledged"** economic theory pertaining to value and price based on productive knowledge, (technological change). Price is the most important originator and regulator of the market system as a whole. The intensity and the magnitude of demand, the supply strategy of the firms, and the efficient allocation of resources are all susceptible to and regulated by the price signals. Thus, economic science urgently needs and requires a new theory of value and price which gives due credit to productive knowledge and is capable of explaining technological change in relation to value generation and price formation both in the industrial and the service sectors. Value-price theory needs to be urgently reconstructed incorporating both productive knowledge and mental labor.

The following section will be an attempt to introduce an alternative growth model. The main assertion is this: all value generation, thus economic growth, has its roots in the intellectual labor of the mind, given the natural resources that are available at that particular time. The new productive knowledge (technology) is embodied in the products that are supplied. In other words, **creative mental labor is the true genesis of all value generation and long term economic growth**.

Value Generation & Growth

A Simple Model Based on Productive Knowledge

Having acknowledged that the productive knowledge (technology) of human labor is both the genesis and the infinite source of continued value generation and prosperity, the remaining sections will attempt to support this hypothesis using a simple model of value generation and growth. It will be assumed that, there is no formal education or training involved and that initially all value is generated by manual labor.

New Technology and Growth

Let us start by examining a simple society consisting of two individuals, Leyla and Maria who possess nothing but their own labor consisting of both mental and manual abilities. In other words, there are no special means of production or of any specific forms of education or training at this time. Assume initially that;

1. Leyla and Maria produce **X** and **Y** goods, respectively; and
2. a working day consists of 10 hours, in the quantities expressed below;
   4 units of X and 2 units of Y

At the end of the day, they exchange their products in accordance with a given consumption level and both have identical preferences. In other words;

Leyla works 10 hours and produces 4 units of X,

And,

Maria works 10 hours and produces 2 units of Y.

The total joint supply a day is $4 X + 2 Y = 20$ hours' work.

Where 1 unit of **Y** is exchanged for 2 units of **X,** i.e.,

$1 Y = 2 X$

The exchange rate seems to be quite fair and rational. Given the consumption level and identical preferences, after the exchange,

Leyla consumes $2 X + 1 Y$ (= 10 hours' manual work

Maria consumes $2 X + 1 Y$ (= 10 hours' manual work).

New Technology and the "Productivity Growth"

A Given Product but introducing a New Method of Production

a- Quantitative Increase

Let us assume that one day one of the producers, say Leyla, utilizing her mental abilities develops a productive knowledge (technology) which enables her to double her daily supply from 4 units to 8 units of **X** within the same 10-hour time-span. To be more specific, Leyla utilizing her intellectual skill, makes a simple tool that increases her productivity of the supply of the "good-**X**". At this stage, some productive knowledge (technological change) has entered the production process (in the shape of a simple tool), although the quantity of hours consumed in the supply of **X** has not changed. **The tool** developed by Leyla is, simply the **knowledge applied through human labor to transform (rearrange) the natural resource**.

Analyzing Leyla's contribution in **quantitative** terms let **Q** denote the total supply, **q** denote each individual producer's supply and **t** the time. The initial joint supply was;

$$Q_t = q^L_t + q^M_t = 4X + 2Y$$

After the introduction of Leyla's intellectual contribution (new technology denoted by **T**), Leyla's output doubles while Maria's remains unchanged;

$$q^L_{t+1} = 2(q^L_t)$$

And

$$q^M_{t+1} = q^M_t$$

As a result of Leyla's productive knowledge (T^L), the total output (**Q**) increases;

$$Q_{t+1} > Q_t \qquad\qquad Q_{t+1} = q^L_{t+1} + q^M_{t+1}$$

The new total supply per day is equal to 8 X + 2 Y which is equal to, in total, 20 hours of manual work + T^L

Though the hours effectively worked (20 hours) have not changed, the community as a whole increases its total wealth due to productivity growth. In other words, Leyla's intellectual contribution has generated growth and increased the community's total wealth. **T** which denotes the technological change (productive knowledge), is not identical with the qualities of the laborer, which will be denoted as L^{α}

b- Increase in Terms of the Market Values and Prices.

Assume that the respective prices, say p_x=3TL and p_y=6TL. The initial total value of output or total income, (TR), would be:

$$TR_t = p_x * q_x + p_y * q_y = 3*4 + 6*2 = 12 + 12 = \textbf{24 TL}$$

After the introduction of the new technology, Leyla's intellectual contribution, the new total income rises to:

$$TR_{t+1} = p_x * q_x + p_y * q_y = 3*8 + 6*2 = 24 + 12 = \textbf{36 TL}$$

The increase in total revenue (ΔTR) equals the increase of Leyla's income (ΔR^L), which is 12TL, while Maria's income (R^M) remains unchanged. The community's total income is now greater;

$$TR_{t+1} > TR_t$$

As is Leyla's income (R^L);

$$R^L_{t+1} > R^L_t$$

While

$$R^M_{t+1} = R^M_t$$

Leyla's income is now twice as much as before due to the technological change that doubled her output per hour worked. The community owes its increased total income to Leyla's mental contribution.

New Technology and the "Exchange Relations"

A given Product but a New Method of Production

What would happen to the other person's (Maria's) exchange relationship in terms of the **relative and absolute prices**, given a sufficient demand?

In regard to the new situation, the exchange relationship will have to change. Previously, there were 2 units of **Y** and 4 units of **X** on the market. Now, there are 2 units of **Y** and 8 units of **X**. What would the new exchange ratio look like?

An Exchange with Relative Prices

Case-1:

Following in the footsteps of the 19th Century economists like Ricardo and Marx, one could argue that it still requires 10 hours work to produce 2 units of **Y** or 8 units of **X**. Equal quantities of labor time are consumed in both cases, and therefore, 1 unit of **Y** should exchange for 4 units of **X** instead of 2, in order to

106

maintain the equality of exchange of the labor-time employed. As a result, at the end of the day, Leyla would be expected to give up 4 units of **X** which equals 5 hours' labor-time for 1 unit of **Y** which also requires 5 hours of labor-time a day. Let **C** denote consumption, $\mathbf{L^\delta}$ the physical labor-power, $\mathbf{L^\alpha}$ the qualified (educated-trained) labor-power and the total labor force **L** be equal to $\mathbf{L^\delta + L^\alpha}$

Leyla's consumption = 1 unit of Y + 4 units of X = 10 hours labor-time

Maria's consumption = 1 unit of Y + 4 units of X = 10 hours labor-time

Total consumption = 2 Y + 8 X = 20 hours' $L^\delta + T^L$

That means;

$$C^L_{t+1} > C^L_t \quad \text{and} \quad C^M_{t+1} > C^M_t$$

If one ignores Leyla's mental labor contribution, exchanging 1 unit of **Y** for 4 units of **X** would seem, at first sight to be an egalitarian exchange. But Leyla as yet has not been rewarded for her intellectual contribution to the common wealth, which has increased the total available supply by 4 units of **X**.

In terms of the hours consumed, the combined employed physical labor-time still consists of 20 hours. But in terms of the initial conditions, the new total output is worth 30 hours of physical labor-time. Maria who's productivity remains constant becomes the main beneficiary of the new exchange relationship based on the "time consumed" approach as she works for 10 hours but consumes 15 hours' worth of output in terms of the initial conditions, while Leyla, producing 20 hours' worth of output in terms of the initial conditions, consumes only 15 hours' worth of output. This would be neither logical nor economically rational from the point of view of further developing a new technology (productive knowledge). The system is unable to provide any incentive for contributing any mental effort.

Case-2:

Assume that after the introduction of the new technology developed by Leyla, which doubled her productivity from 4 to 8 units of X, the initial exchange relationships are maintained. Maria and Leyla still exchange and consume 1 unit of Y and 2 units of X each. But now, Leyla has access to an additional 4 units of X, which she can exchange for another product, say for 2 units of W. Maria still consumes 1 unit of Y and 2 units of X (worth 10 hours output) while Leyla now has 1 unit of Y, 2 units of X and additional 2 units of W at her disposal for her daily consumption, due to her new technology. The total value of Leyla's consumption in terms of labor time consumed has now risen; the initial ten hours' worth of output acquired in her own community plus the value of two units of W acquired from an external source, although the hours she effectively works have not changed.

Leyla's consumption =1 unit Y+2 units X+2 units W = 10 hours' $L^\delta + T^L$

Maria's consumption = 1 unit Y + 2 units X = 10 hours' L^δ

In other words,

$$C^L_{t+1} > C^L_t \quad \text{but} \quad C^M_{t+1} = C^M_t$$

In this case, there is no egalitarian exchange in the Ricardian or Marxist tradition of the equal quantities of manual labor expressed in terms of the time-units consumed. Nevertheless, neither Leyla nor Maria consumes less; in fact, there is an increase in the total consumption due to Leyla's contribution in the form of her productive knowledge. As a result, she is now able to consume more than ever before. This outcome seems both, more logical and economically rational, than in Case-1.

Case-3:

Dismissing Case-1 as being unrealistic and unlikely, let us study a similar but a separate case in which also Maria benefits, such as the service sector which in reality is less dynamic than the industrial sector.

Assume that only 2 units of **X** out of the 4 surplus ones are exchanged for 1 unit of **W**. Leyla would now have 6 units of **X** at her disposal in her own community before entering into the exchange relationship with Maria, and the other member of the community. Assume that Maria after some negotiation somehow "convinces" Leyla to accept a new exchange ratio; say, 1 unit of **Y** for 3 units of **X**. Now, it is not only Leyla who enjoys a greater amount of consumption but also Maria who actually did not make any contribution to the initial condition.

Leyla's consumption=1 unit Y + 3units X + 1 unit W =10 hours' $L^\delta + \delta T^L$

Maria's consumption = 1 unit Y + 3 units X = 10 hours' $L^\delta + \beta T^L$

δ and β denote the parameters for the weights assigned to T.

To put these conclusions differently;

$$C^L_{t+1} > C^L_t \quad \text{as well as} \quad C^M_{t+1} > C^M_t$$

This outcome seems to have a closer resemblance to reality than the two prior cases, because it allows the less productive sectors of the economy (e.g., the service sectors) to benefit overall from the development originating from the "dynamic" sectors. The outcome of Case-3 is, probably, the most realistic and fairest result from the point of view of "distributive social justice".

Exchange with Respect to Market Prices

In the previous section, we studied growth and exchange relationships in terms of relative prices relating to a community consisting of only two individuals.

The market prices in the real world, however, are not determined by relative exchange values. On the supply-side it depends partly on the cost of production (prices of inputs and wages) of each specific product subject to the competitive environment and partly on the magnitude of the demand from the end-users' perspective. For the producer, the market price (exchange-value) of an individual product is expected to be above its objective value, e.g., the cost of its production, if a sustained supply is expected. The upper limit of the market price would be what the market can sustain in regard to the competition and the supply-demand relationships, where the purchasing power and the level of preference play an important role. In order to be able to get a sound insight into actual economic relationships, the emphasis in this section will be on the determination of the individual market price of a single-product enterprise, given a stable supply and demand.

The Sale (Market) Price (P)

Let's assume that there are no costs arising from the "intangible" service activities such as marketing and/or distribution. The price (**p**) reflects the sale price at the factory site which consists of both the costs and the profits. To realize production, the producers combine the material inputs of production (raw materials, components and the means of production, i.e., capital goods) with the services of any labor-power, to produce commodities with specific exchange values. During the process, the enterprise incurs costs called the "costs of production", e.g., payments for the inputs of production. Since producers are driven by a profit motive, the market price of the output supplied is, naturally, expected to exceed the initial costs of its production, in order to compensate the enterprise for the **risks** it has undertaken. The price including the profit is referred to as the "sale price" (**p**) and consists of the following components;

$$p = \text{Costs of production (TC)} + \text{Profits } (\pi)$$

Given the optimum utilization of the productive capacity and the supply-demand stability, and excluding the costs of trading and transportation, how would a technological innovation affect price and income? This can be done by using some data. The intention is not to show how a single price is determined, but to show the impact of a technological change on price and income, cet. par.

An hypothetical case:

Once again let's assume, that the initial supply conditions prevail, i.e., production costs (**TC**) comprise of **LWC** only. In other words, Leyla produces 4 units of **X** at the price of 3 ($p^x_t=3$) and Maria produces 2 units of **Y** at price 6 ($p^y_t=6$). Given the preferences, the income and the consumption pattern, the value of consumption (C_t) for Leyla and Maria, respectively, would be;

$$C^L_t = 1Y + 2X = p^y_t * q^y_t + p_x * q_x = 6*1 + 3*2 = 12 \text{ TL}$$
$$C^M_t = 1Y + 2X = p^y_t * q^y_t + p_x * q_x = 6*1 + 3*2 = 12 \text{ TL}$$

And the total value of income ($\mathbf{TR_t}$):

$$TR_t = C^{L+M}_t = C^L_t + C^M_t = 24 \text{ TL}$$

Case-1: A Given Product a Given Price but a New Production Method

Assume that Leyla doubles her productivity after the introduction of her new technology. Given the demand from external markets, the total value of disposable income for Leyla and Maria ($\mathbf{R^L}$ and $\mathbf{R^M}$) would look like this;

$$R^L_{t+1} > R^L_t \quad \text{while } R^M_{t+1} = R^M_t$$

Because

$$R^L_{t+1} = p_{x+1} * q_{x+1} = 3 * 8 = 24 \text{ TL}$$
$$R^M_{t+1} = p^y_{t+1} * q^y_{t+1} = 6 * 2 = 12 \text{ TL}$$

The community's new total income level is;

$$TR_{t+1} = R^L_{t+1} + R^M_{t+1} = 24 + 12 = 36 \text{ TL} \quad ; \quad TR_{t+1} > TR_t$$

In other words, although there has been no change in price, the community is now richer due to technological productivity growth.

Case-2: Flexible Price and a New Technology

What happens in regard to trade between Leyla and Maria after the technological productivity growth, if Leyla cannot find a third party to trade with at a given price level? The terms of trade will have to deteriorate for Leyla, if the market is to be cleared at the end of the day.

Assume that due to the new technology, the labor time necessary is reduced by 50 percent for the supply of **X,** thus reducing the labor cost of production ($\mathbf{LWC}^L_{t+1}$) by half. The new price will have to be reduced by 50 percent, e.g., $\mathbf{p^x_{t+1}=1.5}$, if the market is to be cleared. The new income level would be expressed as:

$$R^L_{t+1} = 1Y + 4X = p^y_{t+1} * q^y_{t+1} + p^x_{t+1} * q^x_{t+1} = 6*1 + 1.5*4 = 12 \text{ TL}$$
$$R^M_{t+1} = 1Y + 4X = p^y_{t+1} * q^y_{t+1} + p^x_{t+1} * q^x_{t+1} = 6*1 + 1.5*4 = 12 \text{ TL}$$

And total revenue or consumption:

$$TR_{t+1} = R^{L,M}_{t+1} = R^L_{t+1} + R^M_{t+1} = 24 \text{ TL}$$

This "egalitarian" outcome in regard to the Classical approach fails to reward Leyla for her mental contribution adequately and certainly does not reflect the actual transactions in the real world. What is likely to occur is that the price of **X** would settle somewhere between the initial price with a 20 percent profit (3TL) and a break-even price for production, depending on the magnitude of demand.

New Products and New Methods of Production

In the preceding sections, Leyla's productive knowledge, (the new technology), had doubled her productivity of the product-**X**. This sort of growth cannot go on forever. There is a limit to growth unless new products are introduced. **It is the new products**, e.g., new commodities and services that actually **gave rise to the increase in living standards (growth)**, in the long term.

For the sake of simplicity, let's assume that the new products introduced are always accompanied by new production methods. How would an increase in the output of a new product using new production methods, affect the wealth of a community?

The Quantitative "Macro" Growth

Our community consisting of two members, Leyla and Maria, were producing 8 units of **X** and 2 units of **Y**, respectively, after the introduction of a technological change thought up by Leyla. Together, they had;
$$Q_i = q^L_1 + q^M_2 = 8X + 2Y \qquad ; \qquad i = 1,2$$
Assume now that Maria, utilizing her intellectual labor and rearranging the natural resources, produces an "all new" product, say q^M_3 (6 units of **W**). The new total supply, i.e., the wealth, would increase by six units of **W** to:
$$Q_i = q^L_1 + q^M_2 + q^M_3 = 8X + 2Y + 6W; \qquad i = 1,2,3$$
The community is enriched by the quantity of q^M_3. Since there was no previous output of **W** to compare with, one cannot measure the impact of this productivity growth on the price of **W**. All one can say is that there are now 6 entirely new units of **W** at a given price.

The Growth in the Added Value at a "Macro level"

Prices before the introduction of Maria's intellectual contribution were $p_x = 3$TL and $p_y = 6$TL. The total and individual incomes for Leyla and Maria respectively were;
$$R^L_{t+1} = p_{x,t+1} * q_{x,t+1} = 3*8 = 24 \text{ TL}$$
$$R^M_{t+1} = p_{y,t+1} * q_{y,t+1} = 6*2 = 12 \text{ TL}$$
$$TR_{t+1} = R^L_{t+1} + R^M_{t+1} = 24 + 12 = 36 \text{ TL}$$
Maria continues to work 10 hours a day but she is more productive now due to her increased productivity as a result of her intellectual contribution. Say, the new product is sold at price, p_z **5TL** per unit and supplied at the quantity of six

units per sale (q_z 6 W) and costs consist of **LWC** only, as in the previous cases. Given demand at the prevailing price level, the total and individual incomes after the technological change would be:

$$R^L_{t+2} = p_{x,t+2} * q_{x,t+2} = 3*8 = 24 \text{ TL}$$
$$R^M_{t+2} = p_{y,t+2} * q_{y,t+2} + p_{z,y,t+2} * q_{z,y,t+2} = 6*2 + 5*6 = 42 \text{ TL}$$
$$TR_{t+2} = R^L_{t+2} + R^M_{t+2} = 24 + 42 = 66 \text{ TL}$$

which implies that the value of the output per day by Maria has increased from 12 to 42TL, equivalent to the value of q_z.

The Accumulation of Knowledge

Knowledge has been defined as a product of the human mind, the intellectual labor of Homo-sapiens. "Scientific" knowledge tells us that the evolution of the human brain has been going on for tens of thousands of years. Once upon a time the neo-cortex of the brain was much smaller, less developed and less functional. Environmental effects and evolutionary changes over a period of time made the human brain much more productive. One day, human beings learned to make use of flints and bones and used them as hunting weapons, (the hunter's "capital" in Ricardo's thesis). Making simple tools from nature's gifts (flints and bones) to use in hunting or, to assist in physical labor, was a giant step forward and separated humans from other species. The benchmark that distinguished humans from the other species was the application of this "productive" knowledge in transforming (rearranging) natural resources in order to make simple tools which met their basic needs and controlled their immediate environment which made their life easier.

There was at the time, no formal schooling, training, or written material that could be used to pass on this "useful" knowledge to the next generation. There was only the verbal transfer of wisdom by the elderly or by the more experienced members of the community. Nevertheless, mankind continued to benefit from the use of its own mental abilities and kept producing new knowledge that in turn kept improving living conditions and productivity. It took millennia to invent and use very simple tools an extremely slow process when we compare it to the rapid development of our present technology. But these "technologies" were no less important and had a far greater impact on their society than the latest model of computer has on ours today.

Centuries and millenniums past and the human brain continued to produce and accumulate more and more new knowledge, not only to enrich human material standards but also the standards of spiritual life by introducing poetry, music, painting, etc. In time, human beings learned how to pass on this accumulated knowledge to the next generations in the written form, thus facilitating the further and faster development, accumulation and distribution of knowledge. Meanwhile, human beings had also learned to produce products in excess of their immediate needs in order to exchange them for other goods, which further helped to increase the total welfare of the community. By the time the industrial revolution occurred mankind had accumulated sufficient productive knowledge, experience and financial resources to start manufacturing commodities only for exchange. The intellectual contribution of the human brain accelerated productivity growth and the transition to a money economy with enhanced property rights. In our era, not only "tangible" objects like land and material things, but also productive knowledge is subject to proprietary ownership.

Nobody, no matter how brilliant his or her mind is, does **not** acquire knowledge in the form of "manna from heaven". The present pool of productive knowledge is accumulated through thousands of years and is a common heritage. In the absence of an accumulated productive knowledge, there would be no products to be exchanged. In the present era, acquiring the appropriate knowledge for production through schooling and training is very important in order to sustain as well as to improve the wealth of all nations. There is a close correlation between a knowledgeable labor force and a per capita productivity level, because the better and longer the education and training it will improve the productivity of the individual as well as of the nation as a whole. If it were possible by waving a magic wand to transfer, in one night, all the production facilities in Germany to Turkey, the total output would diminish both in terms of quality and quantity simply because the Turkish labor force is not as well educated or trained as the German labor force. In other words, the quality of the labor force is a very important factor in the productivity and prosperity of any nation.

However, although necessary, it is not simply sufficient to have a well-educated labor force. Countries like Russia and Ukraine have a more-educated labor force than their US counterpart in terms of hours of education invested (the so called "human capital"). Yet, the per capita added value of a Russian worker is much lower than the US-worker. That is because the institutional and cultural settings in the sense of the competitive firms, the contemporary entrepreneurs or managers, the government's economic policies, and the technical, financial and legal infrastructure, are all important factors for the efficient functioning and success of the system as a whole. The education and training of the appropriate labor force is a long process, but the development of an appropriate institutional and cultural environment might take even longer and be more troublesome.

Some Ethical Questions

In our simple community with two producers, Leyla's contribution was a new technology. It was a small step, a tiny drop in the ocean, but the available productive knowledge has been accumulated for millennia through such marginal contributions. Now we have access to a grand pool of knowledge to dip into which is, in fact, the common heritage of all mankind. Existing knowledge is being further developed by the creative contributions of inventors who are being educated by existing knowledge using public resources. In a sense, new productive knowledge is not an entirely new created exclusively by some person or firm from scratch. All new productive knowledge is a small even minimal contribution to pool of the accumulated knowledge of the ages. No technology emerges as if it were manna from heaven, except in Solow's model.

In some circumstances, if some person or a firm acquires a new patent for a new technology, how exclusive should these patent rights be? To what degree should these new ideas belong to an individual marginal contributor? To what extent is society entitled to new patented knowledge, as it is the provider of the educational facilities and the scientific and technological infrastructure? And finally, to what extent are the past contributors to the grand pool of human knowledge entitled to the rights of the patented product?

Concluding Remarks

Given nature's gifts to humankind (resources), a very simple growth model as has been suggested in this paper clearly suggests that the **productive knowledge** (technology), which is a product of the human mind (mental labor), is **the genesis** as well as the **ever continuing source** of all the prosperity generated by man, e.g., **value generation and growth**. The initial inputs of all production are natural resources and the labor of man. In all societies, all the inputs and outputs of any production process are, in principle, the same; labor and the transformed or rearranged natural resources. As we have seen in this paper, **"technology creating"** as well as **"technology using"** labor-power is the most important ingredient of all output, in contemporary economies. The higher the qualities of the laborers, the higher would be the expected output and the level of wealth. **"Technology creating labor-power"** deserves special attention as it is the source of all knowledge of production and long term growth.

The major difference between ancient and contemporary societies is that we now have access to a tremendous amount of "accumulated knowledge" so we have the means of production and we can make use of this "accumulated productive knowledge" to assist us in our continual search to produce new products. In other words, at present there is more productive knowledge, which means more ability to produce and more products for us to consume.

Technological change is one of the key terms for long term economic growth. The other term is the quality of human resources, especially in the developing countries. The institutional infrastructure, social values, traditions as well as consumer habits in different countries may foster or be impediments to economic growth.

The citizens of the less developed countries (LDCs) enjoy much less prosperity than the citizens of the developed countries (DCs) due to their much lower per capita productivity. But, there is a great potential for the growth in technological productivity even with the present technologies. That means that even without any technological change there is a great long term potential for a rise in global economic growth from LDCs. To be more specific, the global economy can continue to grow by transferring available technologies to the LDCs but only if restrictive clauses can be eliminated (see Gürak;1990). It is also of vital importance to educate and train the labor-force in the efficient use of the available technology, reorganize and restructure the institutional setting and to pursue more appropriate economic policies. Such steps would not only help to reduce disparities among nations but also may cure many global social and political evils.

In conclusion, given the natural resources and the level of our existing knowledge, the **creative mind of the labor force (mental labor), is the only value-producing source, in the long term** of all the past, present and future prosperity. To put it in William Petty's words;

"... labor is the father of (material) wealth, the earth is its mother." (in Marx,1976,Vol. I,pp.133-134)

The prosperity of mankind has one unique source;
"Knowledgeable and creative human beings..."

References

Baumol, W.J.-
McLennan,K. (eds.) 1985 *Productivity Growth and*
US Competitiveness
Oxford University Press, New York.

Becker, G.S.　　1975 *Human Capital.*
National Bureau of Economic Research, N.Y.

Blaug, M.　　1980 *The Methodology Of Economics.*
Cambridge Uni. Press, Cambridge.

----- " -----　　1990 *The History of Economic Thought.*
Edward Elgar Pub.. Ltd., Hants.

Bohman, R.S.　　1990 "Smith, Mill & Marshall on Human Capital Formation
History Of Political Economy, Vol. 22:2
(Brookings Papers on Economic Activity)

Crawford, R.　　1991 *In the Era of Human Capital.*
Harper Business, New York

Drucker, P.F.　　1981 *Toward The Next Economics*
Harper & Row Publ., New York.

---- " ----　　1993 *Yeni Gerçekler*
İş Bankası Kültür Yayınları No: 315

---- " ----　　1995 *Gelecek İçin Yönetim.* (Managing For Future)
İş Bankası Kültür Yayınları No: 327

Grossman, G.M.-
E. Helpman　　1991 *Innovation and Growth.*
MIT-Press, Cambridge.

Gürak, H.　　1990 *Transfer of Technology*
Lic. Thesis, University of Lund. Sweden (Unp.)

---- " ----　　1993 *An Alternative Price Theory*
Post-doctoral Thesis (Unpublished).

---- " ----　　1999 "On Productivity Growth"
YK-Economic Review, Dec., Istanbul

---- " ----　　2000 "Verimlilik Artısları" (Productivity Growth)
Verimlilik Dergisi, May-June, MPM, Ankara

Hausman, D.M.　　1981 *Capital, Profits and Prices.*
Columbia Uni. Press, New York.

Hicks, J.R.　　1965 *Capital And Growth.*
Oxford Uni. Press, London.

| ----- " ----- | 1979 | *Causality in Economics.* |
| | | Basic Blackwell, Oxford. |

----- " -----   1983   *Classics and Moderns.*
Basil Blackwell Publ., Oxford.

Hume, D.   1986   *Insan Zihni*
Milli Egitim Basimevi, Istanbul.

Jones, G.   1971   The Role of Science and Technology in
Developing Countries. Oxford Uni. Press.

Kuhn, T.S.   1982   *Bilimsel Devrimin Yapisi.*
(The structure of scientific revolutions)
Alan Yayincilik, Istanbul.

Lucas, R.   1988   "On The Mechanics Of Economic Development"
Journal Of Monetary Ec., July, 1988,342

Mankiw, G.   1995   *The Growth Of Nations*
Brookings Papers on Economic Activity

Marshall, A.   1990   *Principle of Economics.*
Macmillan And Co., London.

Marx, K.   1976   *Capital,* Vol. I
Penguin Books.

----- " -----   1977   *Capital,* Vol. II
Lawrance & Wishart, London.

----- " -----   1981   *Capital.* Vol. III
Penguin Books.

Meek, R.   1973   Studies in the Labor Theory of Value:
From Smith to Ricardo
Lawrance and Wishart, London, 2nd Ed.

Mill, J.S.   1986   *Faydacilik (Utilitarianism)*
Milli Egitim Basimevi, Istanbul.

McIntyre, J.R.-
D.S. Papp,(Eds.)   1986   The Political Economy of
International Technology Transfer.
Quorum Book, New York.

Nelson, R.   1994   What Has Been The Matter With Neoclassical
Growth Theory ? in Silverberg-Soete (1994 Eds)

Ricardo, D.   1990   *On The Principles Of Political Economy & Taxation.*
Cambridge University Press.

Romer, P.M.   1990   "Endogenous Technological Change"
Journal Of Political Economy, Vol.98, Oct.

---- " ----	1993	"Economic Growth", in D.R.Henderson (Ed.) ***The Fortune Encyclopedia Of Economics***, Time-Warner Books, New York.
---- " ----	1994	"Beyond Classical & Keynesian Macroeconomic Policy" ***Policy Options***, July-August.
Schultz, T.W.	1980	*Investing in People.* University of California Press
Schumpeter, J.A.	1954	*History of Economic Analysis.* Oxford Uni. Press, New York.
---- " ----	1959	*The Theory Of Economic Development* Harvard Uni. Press, Cambridge.
---- " ----	1970	*Capitalism, Socialism And Democracy.* Unwin University Books, London.
Silverberg, G. – L. Soete (Eds)	1994	The Economics Of Growth And Technical Change. Edward Elgar Publ.
Smith, A.	1976	An Inquiry Into The Nature And Causes Of The Wealth Of Nations, Vol. 1 & 2.
Solow, R.M.	1988	*Growth Theory: an exposition.* Oxford Uni. Press, New York.
Toffler, A.	1992	*Yeni Güçler - Yeni Soklar. (Powershift)* Altin Kitaplar, Istanbul
UNCTAD	1983	Draft international code of conduct on the transfer of technology. TD/CODE , TOT/41

4- AN ALTERNATIVE GROWTH MODEL

Introduction

The content of this paper is acquired from the book titled "Economic Growth and Global Economy" (Chapter-5). In that book, the analysis of economic growth commences with the critical analysis of Classical period economists like A. Smith, Ricardo and Marx, and continues with the evaluation of works of some well-known 20th century economists such as Marshall, Keynes, Schumpeter, Harrod-Domar, and ends with a critical presentation of the works of the neoclassical heritage, including the so-called "endogenous growth" theories. An attempt shall be made, in this work, to construct an alternative theory of growth in order to make a further contribution to the analysis of economic growth.

It is well known fact that the relationship between the labor(-er) and value-production, was a top economic priority in the research conducted by the economists of Classical period. At that time, technological progress used to play an important role in their dynamic analysis and was treated as an endogenous factor. But, in spite of the important role assigned to it in their economic analysis, the Classical economists failed to construct any satisfactory growth models which demonstrated the inter-relation between technological progress and economic growth.

After the 1870s, dynamic economic growth analysis began to be replaced by the "static equilibrium" analysis of the Marginalist and Neoclassical doctrines. The aim of the new doctrines was to find new methods which could bring economies into equilibrium. Leaving aside the attempts of Schumpeter in the 1930s and 1940s which re-emphasized the importance of technological advances in economic growth, the dominant economic growth models of this neoclassical heritage completely ignored this technological aspect, until the appearance of Solow's contributions in the 1950s. In other words, the economists of the neoclassical heritage according to the models they proposed had no idea at all of the impact of technological changes and how they affected the course of economic progress.

The advocates of the neoclassical doctrine "rediscovered" technological progress and the fundamental role it played, thanks to the work of Solow. Using an analogy, they were very happy at finding the dog, which they neglectfully lost. After his "striking" rediscovery, Solow was awarded the Nobel Prize. However, Solow had some problems with the origin of technological progress and could not explain how it emerged from within the system. But, he found an ingenious solution to this problem by declaring that technological progress was "exogenous" to the system. They were being produced externally by non-economic factors and "fell" into the economic system like "manna from heaven", and made amazing impacts on economic growth.

Despite all the shortcomings of the new theory, Solow's "rediscovery" engendered a renewed interest in the relationship between dynamic growth and technological progress among economists. Later, economists also "rediscovered" the crucial role played by the "qualities" of the laborer (human capital) in this process. In the 1980s, new equilibrium models called "endogenous economic growth models" began to emerge, which considered technological progress as deliberate and conscious consequence of economic decisions. However, all these models had many serious shortcomings in fully explaining the actual process of economic growth. These particular shortcomings have been discussed in the book entitled "Economic Growth and the Global Economy", by Gürak (2006). At present there does not appear to be any single economic growth model about which economists have displayed any consensus. There seems still to be a need and scope for further development in the field of long term economic growth analysis. The shortcomings of these prevailing theories are the driving force behind this present work.

It is the intention of this work to present an "alternative" long term growth model based on a key concept namely the "qualifications of the laborer". We encounter this concept, the "qualifications of the laborer, in every aspect and at every level when we undertake any long term growth analysis. To put it more specifically, given the gifts of nature, "creative mental labor" appears to be both the genesis of and the reason for the continuity of all long term economic growth. Technological advance, which is the indispensable ingredient of all long term growth, is, in fact, the product of "creative mental labor". The efficient use of these technologies is also an essential factor, but, its efficiency is also related to "mental labor". The qualities of this labor(-er) can be divided into two general groups:

1. Technology-producing labor;
2. Technology-using labor.

The technology-producing "creative mental labor" precedes the latter in importance in the formation of long term economic growth.

The probability of the acceptance of an alternative growth analysis as presented in this work is not high, especially by the proponents of the neoclassical doctrine who rigidly adhere to their equilibrium analysis. Furthermore, some economists may think: Why should we bother with an alternative theory of some unknown economist while there are plenty of works by well-known economists? They might have a point. But, if these economists decide to be more open-minded and for a short time leave aside their "precious" devotion to the doctrines to which they are committed and provide this alternative theory with, say 10 percent, of the degree of the objectivity and tolerance they gave to the well-known Solow's articles, which in fact study only short term growth analysis and embody many serious shortcomings, they may actually encounter some "logical, and consistent explanations" for long term economic growth.

Labor, Technological Progress & Growth

Retrospectively, one observes that the economic growth analyses of the older "Classical" economists were quantitatively and qualitatively less sophisticated than are used today. But, nevertheless, their analyses appear to be more useful in grasping actual economic relationships when compared to the contemporary neoclassical analysis. For instance, the crucial factor of growth, "technological progress" which was "rediscovered" by the neoclassical economists in the 1950s, was already in existence in the analysis of the Classical economists as an "endogenous" factor. The same applies for the concept and role of "human capital", or rather the "qualifications of laborer", which was "also" rediscovered in the 1950s by the Neoclassical economists. However, more than 200 years ago, Adam Smith had written about the importance of the education of the labor force in regard to output and advocated that the children of poor families should have access to free education. At that time, there was no talk of a social state at all. Although technological progress was regarded as an "endogenous" process, being driven by internal forces, the Classical period economists had, unfortunately, failed to construct a coherent theory on long term sustainable economic growth based on creative mental labor and technological progress.

After the 1870s, the ideological process of transforming what then was known as the "political economy" was initiated in order to pave the way for a "purely scientific" economic analysis, just like the one's found in the natural sciences e.g., astronomy. These new methods of analysis which were free of any human or subjective values gradually increased with ever growing sophistication of the mathematical models. But this caused many serious deviations from actual economic events and led economists to favor a "utopian" approach in dealing with economic relationships. All historical developments, institutional settings, cultural-political relations and personal values had suddenly lost all their importance. Instead, the new emphasis was on how to engender "static equilibrium". Yet, "static equilibrium" has never actually taken place and, in all actuality, never will.

As a result, over a period of time the neoclassical ideology gained increasing attention and respect and acquired global recognition. Meanwhile, around 1938 the English economist Keynes succeeded in shaking the foundations of the neoclassical theory in regard to such basic issues as full-employment and an automatic return to a state of equilibrium if everything was left solely to market forces. But the fundamental concept of long term equilibrium remained unchallenged. In fact, Keynes was attempting to outline a path to long term static equilibrium. Since the main focus was on acquiring equilibrium, the Keynesian theory had nothing at all to say about long term growth. But then what? Does such a short term equilibrium theory, which has nothing to say about human capital and technological progress, deserve to be acknowledged as globally applicable?

Actually, it seems to be more appropriate to regard the Keynes' analysis as a "different version" of the neoclassical "static equilibrium theory". Keynes had never objected to the concept of static equilibrium and had nothing to say about long term developments. He had mentioned labor, but never made any reference to the qualifications of that labor, that is "human capital". He saw no correlation between labor and technological progress. In his analysis his theoretical criticism of the neoclassical doctrine was not unjustified in regard to employment and disequilibrium. But Keynes had "nothing new" to say about the political measures which intervene in the market to reduce unemployment. At the time of the publication of his book, the states in Europe and USA were already implementing some policy measures to tackle the widespread unemployment problem. In other words, the so called "Keynesian policies" which proposed to increase employment and output were already being practiced in several parts of the world, long before Keynes mentioned them in his book. Some economists claim that the "Golden Age" in the application of Keynesian policies was in the post Second World War era until the time of the oil crises between 1945 and 1973. But this is not correct either, for that period was not characterized by any unemployment, but, on the contrary, by a shortage of laborers, which led to the mass "imports" of laborers into Europe, Australia and New Zealand.

In fact, in order to be fair to Keynes, it would be proper in making a distinction between his theoretical thoughts and those of the so-called "Keynesians". Because, there are, at times, considerable differences between them. As mentioned above, at the time Keynes' book was published, some countries were already applying the so-called "Keynesian policies", enabling government intervention to eliminate unemployment. Keynes' book was an important theoretical contribution to the neoclassical equilibrium analysis; but its practical contribution to a real economy is not that certain. In short, the contributions of Keynes and the so-called "Keynesians" ought to be analyzed and evaluated separately on their own merits.

As to the prevailing neoclassical ideology; in spite of the fact that it basically preserves the "status quo" and does not appear to have diverged from its original path towards a state of equilibrium, it appears to have been subject to serious setbacks from the proponents of ideologies such as those of Solow et. al., with their "technological progress" and "qualified labor" approaches. Until the study of the causes of growth by Solow in the 1950s, the focus of economic analysis highlighted business cycles and the restoration of an "equilibrium", which, in fact, never existed. From the 1870s on until the 1950s, for about 80 years, the level of output was assumed to be determined by the level of employment of the two factors of production, labor and capital. Rediscovery of the effect of technological progress on growth of had caused the production function to change, followed by another rediscovery, that of "human capital". In other words, the original production function of the Neoclassical doctrine, **P=f(K,L),** first became **P=f(K,L,A),** and then **P=f(K,L,A,H).** But, L and H are in principle, the same thing, namely two sides of the same coin. Moreover, "A" originates from "creative mental labor" i.e., "L" and the knowledge created by mental labor is embodied in a physical form in the means of production. The neoclassical analyses inevitably led to many analytical misconceptions, misperceptions and miscalculations. Since the 1980s, many attempt have been made to indigenize the technological progresses and human capital in the so called "endogenous growth models". A lot of progress has been made, but there still seems to be scope and a need for further development in the theory of growth.

The Genesis of Growth: The Qualifications of the Labor Force

The purpose of this work is to make a contribution to the theory of growth by presenting a "simple" growth model based on the concepts of "creative mental labor" and "technological development". The gifts of the nature, e.g., raw-materials and their market values are assumed to be a given. According to this hypothesis, the genesis of all the value added to the prevailing market values of raw-materials, thus the continuing source of all output growth and a nation's riches, is laborer, or rather, the level of qualification of the laborer. With the application of technology, which is the result of mental labor, in the production process, raw-materials (gifts of nature) are transformed into "useful" products such as tools, intermediary inputs or consumption goods. To put it another way, the technology produced by the laborer is used to produce an output either for further production or final consumption. When new technologies are embodied in the tools of production, they may either help to increase the "per unit time productivity" of employees, or introduce new products. Thus, the concept of the "marginal productivity" of capital (-goods) is nothing but a fallacy.

In the subsequent section, the concept of labor(-er) will be categorized under four headings, in order to get a clearer insight into its contribution to the growth process:

1- Physical labor (L^b).
2- "Creative" mental labor (L^y).
3- Technology using labor of varying quality levels (L^k).
4- Labor's level of qualification consisting of L^b, L^y and L^{k_i}, (L^n or simply L).

Physical labor refers to all kinds of basic physical activities such as walking, drinking, holding, etc. Such activities are similar to those made by other living beings in order to survive in their environment. Such activities can be initiated by basic instincts. But, nevertheless, the control center of every type of activity is the brain, and in the absence of these mental directives, living beings could not survive. Even the most basic activities are initiated by instructions from brain. Therefore, there is always some degree of mental contribution involved in every stage of existence. The concept of physical labor, in our case, simply means the carrying out of these sets of instructions, which result in the basic co-ordination of the human physical system.

"Creative" or "productive" labor is the source of all the added value accrued from any creative activities or changes, which involves those activities beyond the basic sets of instructions sent from the brain governing purely physiological activity. In modern societies, creative labor is, in general, employed in R&D departments in search of new ideas. Research funds are normally employed to finance the creation and development of either" new products" or "new production methods" which in turn are used to produce the available goods or services at a lower cost. Naturally, it is not only the highly educated, but also those with a lower level of formal education who are able to contribute to the creation of new ideas. One way or another, the new ideas or the knowledge required to raise the productivity per unit time employed and the standard of living is always a result of the creative abilities of human beings. Accordingly, all technological developments necessary to increase both individual and total productivity also stem from the creative mind.

Technology using labor of varying quality levels: Those who produce new technologies and those who employ these technologies are, in general, not the same laborers. There is always a need for qualified labor to efficiently employ the existing technologies. In other words, the degree of utility of a technology depends on the qualities of existing labor force. For instance, if the labor force is not properly equipped with the knowledge necessary for efficient production, it would be impossible to produce, say, airplanes or automobiles. Therefore, the qualities of the labor force are important for the efficient production and the total wealth of society, along with laborers who are able to use their creative mental abilities.

Labor's level of qualification (L^n): As mentioned before, nowhere, there is a labor-force without some degree of qualification (i.e. without any form of education, training, skills) i.e. a labor force consisting only of purely physical laborers. In real life, every individual's labor possesses some degree of qualification that is referred to as human capital. Therefore, it would be a serious error to label the labor force in terms of the two artificial categories presently being used i.e. as being "qualified" and "unqualified". This error will not be repeated in this work, unless in appropriate and exceptional cases. The descriptor " L^n " or simply "**L**", will refer to both the physical as well as the mental abilities of the labor force.

There are four basic factors determining the level of qualifications of labor and its productivity:

    1- The individual's "natural" capabilities and talents.
    2- The general level of the knowledge in society.
    3- Formal and informal education.
    4- Learning-by-doing.
    5- Experience.

Any increase in any one of the factors mentioned above would in turn raise the qualifications of the labor force. Every capable individual enjoys a certain level of ability. In other words, every individual possesses a certain degree of the five factors stated above. At the least, in a modern society some capabilities would be acquired through informal education from the family and others from environmental factors. In advanced societies learning would be more visible as almost every individual attends the school for at least 12 years. Therefore, the claim that every individual possesses some degree of qualification (human capital) is not easy to challenge. In fact, it would be more appropriate to talk about the different quality levels of the laborers rather than drawing a distinct line between a qualified and an unqualified labor-force.

There is a close relationship between the level of a laborer's qualifications and the standard of education in a particular country. The higher the level of qualification of the labor force, the higher the expected individual or total level of wealth would be. On the other hand, regardless of the level of the individual's capabilities, if the nation's general development level is below the global contemporary level, productivity per unit time employed would be expected to be lower than the global standard. That is to say, the level of the quality of the labor force and the productivity of a nation is closely related to the general level of the accumulated knowledge and the quality of the labor force. As mentioned before, creative mental labor is the genesis of all the added value and accumulated wealth, but the level of productivity (the efficient application of technologies in production), depends on the quality level of the labor force.

Some of the productivity growth may be attributed to practice at work. In his third model, Lucas (1988) claims that the necessary capabilities for production are acquired through the "learning-by-doing" process. The quality of work done increases as the hours spent at work increases, given the same technology. Workers are so specialized at practicing the job that over time the cost of production is expected to fall while and output to increase. Thus, per unit time productivity would increase with increasing practice at work.

Experience is a concept that has a wider meaning than simply "learning-by-doing". The impact of experience, the importance of which seems to be underestimated, is also significant in the development of mental capability and thus to value generation. For instance, a doctor or a nurse may learn surgical procedure step by step by learning-by-doing with a given technology; experience would make them more efficient in terms of the general issues related to simple surgical procedures. Similarly, the contributions of an experienced person in making a decision on a strategically important issue can be extremely vital. More experienced individuals are more likely to make wiser and less faulty decisions, which may have a great impact on their present and future work. In a similar fashion, a more experienced teacher or security officer who has learned the job through the "learning-by-doing" method is likely to be more productive at work after several years of experience.

Is it possible to measure the "Qualification Level of the Laborer" accurately?

Is it possible to measure the mental labor or the qualifications of the laborer, which are of crucial importance in regard to output as well as output growth? If so, what should be the proper method of measurement? Would the outcome reflect the facts accurately?

The issue of measurement bears a great importance for many economists. In fact, according to many economists, if something is not measurable, it cannot be "scientific". The desire to measure is logical; but to expect to achieve definite and repeatable results in the social sciences is illogical. Five factors were put forward as the factors which can affect the level of qualification of the labor force. Mental labor was considered to be the most important factor as it is the foundation of and the eternal source of long term growth. None of these factors can be measured accurately in order to acquire precise results.

Let us mention again these five factors one at a time: In regard to the first, individuals are endowed with different kinds and degrees of **natural talents**. Some individuals possess talents for sports while some are talented in arts. And there is no technique now nor will there probably be in the future, that could be used to measure such a diverse and multi-level concept such as "talent" in an accurate and universally accepted manner.

The "general" knowledge level varies from country to country. It would not be surprising to find that there is a great gap in the amount of accumulated knowledge between Sweden and Ghana or Pakistan. Sweden has been closely following scientific and technological developments in the past 200 hundred years, while at the same time attempting to establish an appropriate infra-structure. Nowadays, Sweden has many globally competitive firms employing the most advanced technologies, while Ghana and Pakistan seem to be latecom-ers in these fields. Due to the prevailing conditions in their countries, individuals who grow up in those countries encounter different environments and develop-ment levels. Swedish citizens enjoy a wide range of facilities which they can use to learn and benefit from these contemporary advanced technologies, while in-dividuals in the other countries are not even aware of most of these develop-ments. There is no method to measure such divergences and consequently or to make accurate comparisons of their impact on their respective societies.

Formal-informal education and training: To measure the amount of hu-man capital, i.e., the qualifications of the labor force, some economists use the number of school-years attended. Certainly, this can be a method of measure-ment; but it is definitely not an accurate one. For a period of education lasting 12 years there can be and are considerable divergences in the quality of education and training between different countries. It is also a well-known fact that the quality of formal education or training may vary considerably among the schools in a single country, especially in the developing world. Therefore, the measurement of the human capital, (qualifications of the laborer), based on the "formal school-years attended" criterion can never be accurate.

In regard to the measurement of informal education or training, the solution is no less cumbersome. There are no "school years" to count, nor is there any obvious reason to suppose that informal education achieves better or worse re-sults than the former.

Learning-by-doing & experience: The measurement problem is not any better than in the previous cases. There is no available method to measure the "practice" or "experience" level of individuals accurately. Each individual possesses varying degrees of knowledge and skills, which influences the accumulation of learning-by-doing or experience differently.

Let us recall the initial question in light of all these facts: Is it likely that the level of the qualifications of the laborer which are the source of continual long term growth can be measured accurately at all? In order to give a positive answer to this question, one has to be either naïve or optimistic. The best outcome would involve measuring "probabilities" or "tendencies".

Productive Factors and Production Factors (Inputs)

There were two main factors of production considered by the orthodox economic theories; viz. labor (**L**) and capital (**K**). Due to the developments in economic theories since the 1950s, two factors are now added to the original two; namely technological change, (**A**), and human capital (**H**). In this section of this work, the reader will be presented with a different approach to these issues which differs from those prevalent at this time. This approach claims that there are only "**two productive factors**" of production, but "**many production factors**".

Productive factors:

1- Labor(-er) (L) (physical as well as mental).

2- Nature (the entire ecological system)

Factors (Inputs) of Production

The concept "factors of production", as used in this work, differs from the concept used in the orthodox equilibrium theories. In this work all required inputs of production are the factors of production. For instance, along with labor and capital goods, all raw-materials, the energy used, buildings, tools, in short, every item necessary for a required output are factors (inputs) of production. In contrast to the orthodox equilibrium theories, capital goods are not assumed to be productive; but on the contrary, they are used to increase the productivity of the laborer employed in production. The factors (inputs) of production are:

- Labor(-er).
- Raw-materials.
- Intermediaries (semi-finished goods).
- Energy inputs, water, etc.
- Capital goods (machinery, tools).
- Consultancy services.
- Post-production marketing and sale efforts.
- Transport-insurance.

128

- Management.
- And all other inputs required for the particular output.

Factors (inputs) of production can be subdivided into two broad groups:
    1- Labor(-er) (L).
    2- Other inputs (X_i).

"Productive" Factors and Value-Creation

There are only two "productive" factors; nature and labor. Nature is productive in the sense that it is capable of supplying products with "use-value" without any external intervention. These products range from directly consumable products such as vegetables and fruits to the basic inputs of production which in turn are transformed by labor. The productivity of the nature is closely associated with environmental conditions. Nature, in modern societies generally, does not supply products, which are directly consumable. In order to be consumable contemporary products have to be "**transformed**" into "useful" products by some form of labor. It is only after being processed by labor these supplies from nature are transformed into products with an "exchange-value".

The labor-time spent could range, from a "simple" labor-time, say transporting the apples from a garden to the market place, to a more "complex" labor-time requiring higher qualifications in transforming nature's products into semi-finished or finished products. For instance, the raw form of a chair is the tree, and it is transformed by labor into a useful product with an "exchange-value". It is a primary law of physics that nothing in nature disappears completely and nothing is created without using some form of available input; natural supplies only change forms through labor. That is to say, nature supplies the basic inputs of all output and laborer converts them into the other forms demanded by the consumer. Assuming that the inputs of nature are a given, the source of all "use-value" and "exchange-value" is generated by laborers.

Following this line of reasoning, an attempt will be made below to construct a simple growth model based on labor and laborer. It will not be the aim of this simple model to give an exact account of actual complex economic relations. But, rather, it could be used as a precursor to pave the way for more realistic models in the future. Therefore, a simple model will be sufficient to serve our present purposes. Because, the main purpose of the simple model is to show that the original source of all created value and technological innovation is labor, or, more specifically, "**creative mental labor**". Therefore, the reader is asked to bear in mind the mental labor aspect of the model all times.

Some basic assumptions

1- The determining factor of long term growth is technological progress, which is a product of creative mental labor, (L^y).

2- Nature, one of the two "productive factors", supplies the necessary inputs for production, while laborer transforms them into useful products.

3- L^n or simply L, denotes the average labor force endowed with a certain level of qualification (human capital).

4- The level of general knowledge and development in a country determines the general level of qualification of the labor force, L^n, which, in its turn, determines the level of the labor force's productivity. L^n and includes, in addition to physical labor L^b, the mental labor L^y, which in turn creates the new technologies and a higher qualified labor force, L^k, using the new technology in production.

5- Let us denote the qualifications of average laborer in Turkey by L^n_{TR}, and in EU by L^n_{EU}. The present situation due to the differences in qualification level can be shown as $L^n_{EU} > L^n_{TR}$; or, simply, as $L_{EU} > L_{TR}$.

6- There are two countries or producers (TR,EU) with an equal quantity of labor-force. Further assume that the quantity of "physical labor-time" spent in one day is equal in both countries ($L^b_{TR}=L^b_{EU}$). Under these circumstances, the wage rate in both cases should be equal ($w_{TR}=w_{EU}$).

However, the situation described above will change as soon as we take into account the differences in qualifications of the two labor forces. Assume that the average qualifications of labor force in the EU are better than in TR ($L^n_{EU} > L^n_{TR}$). Naturally, the labor force in the EU would be more productive than the labor force in TR and enjoy a higher wage rate ($w_{EU}>w_{TR}$). The difference between the qualifications of the labor forces is due to a relatively higher degree of accumulated knowledge in the EU, better educated and trained labor force, and the technological as well as the institutional development level.

Output, Exchange & Distribution in a "Stationary" Economy

Firstly, we shall consider a stationary economy in which "**only physical labor**" is employed and no growth takes place. The purpose is to analyze the barter-exchange relationships, along with the individual and the total consumption level, using the "physical labor-time employed" approach.

The qualified laborer will be included in the following sections and its impact on growth will be analyzed. The structure of the model facilitates the study of income distribution together with growth.

Assumptions:

- There are only two producers and two consumers, Leyla, (**L**) and Maria, (**M**).
- Only two products are produced and consumed; X_1 and X_2.
- Consumer preferences are the same.
- No accumulation. All output is consumed on the same day of production.
- No money. Barter-exchange takes place.
- Only "physical labor" is used in production, L^b.

Since "creative" mental labor, L^y, has not yet been introduced, there is no new technology (**A**) developed nor any "means" of production (capital-goods) **K**, has been produced. Thus, there is no need for any qualified labor, L^k, for the efficient employment of any technology.

So the function is:

$$Q = f(L^b_L, L^b_M)$$

L^b_L, denotes Leyla's, and L^b_M Maria's physical labor. Initially, both Leyla and Maria enjoy "the same level of qualification" that of simple "physical labor" $(L^b_L = L^b_M)$. Each of them work 10 hours a day and produce two different products (X_1 and X_2). Leyla's daily output is 4 units of X_1 that of Maria is 2 units of X_2, and both have identical tastes and preferences. At the end of each day, they exchange products worth 5 hours of labor-time ($2 X_1 = 1 X_2$). The outcome is:

Leyla's output 4 X_1	10 hours / day	
Maria's output 2 X_2	10 hours / day	
Total output $\quad Q^T = q^L + q^M = 4 X_1 + 2 X_2 = 20$ hours / day		(1)
Leyla's consumption $\quad C^L = 2 X_1 + 1 X_2$		(2)
Maria's consumption $\quad C^M = 2 X_1 + 1 X_2$		(3)
Total consumption $\quad C_t^{L,M} = 4 X_1 + 2 X_2$		(4)

Both, Leyla and Maria, spend equal quantities of labor-time and, as a result, consume equal quantities. The exchange is "fair" in terms of labor-time employed and both enjoy an equal quantity of use.

In the absence of any "mental" labor contribution, which means there are no technological innovations, so no growth would take place, because the production capacity and tastes are as stated. The existing system is capable of only maintaining the status quo of any existing production and exchange relations. Equilibrium exists but there is no growth.

For growth to take place, both the output and consumption have to increase. For output to increase there is need for "creative mental labor" that means, new technologies have to be introduced. There has to be either innovation introducing **a new production method for a "given" product**, or **entirely new products with new production methods**. In the following models, we shall assume that, given the product, **a new production method** increases output.

Barter-exchange and Growth: 1-a

Technological-productivity growth

A Given Product using a New Production method

Additional assumptions:

1- By utilizing her "mental" abilities, Leyla increases the productivity of her labor-time employed in production. The reason for this "improved" mental ability could be the education or training she obtained, or her natural talents, or her experience. Leyla's labor now is no longer purely L^b, but L^n. With the assistance of her "creative mental labor", L^y, Leyla introduces a new technology, which increases the productivity of her labor-time employed.

2- With the introduction of the new technology, the need for new labor skills arises. In other words, for the efficient employment of new technology, the skills of **L** have to improve.

3- Both, economic efficiency (**EE**) and technical efficiency (**TE**) are assumed to be at an optimum level.

4- Supply and demand is in balance. Every additional item of output is consumed, but the exchange-relationships will have to change.

Leyla increases her daily output from 4 X_1 to 8 X_1 with the employment of the "new" technology developed by her mental ability. Initially, Leyla's labor had no qualification level; that is $L^n_{L,t} = 0$.

But now;
$$L^n_{L,t+1} > L^n_{L,t}$$

And;
$$q^L_{t+1} > q^L_t$$

For Maria, the initial conditions are still valid.
$$L^b_{M,t+1} = L^b_{M,t}$$

And;
$$q^M_{t+1} = q^M_t$$

The new total production function is:

132

$$Q = f(L^n_L; L^b_M) \qquad (5)$$

The new technology (**A**) developed by Leyla's "creative mental labor" (**L**^{y}) is embodied in a material form in the means of production and help to increase her productivity.

Since preferences and working-hours have not changed and there is no third party with whom to enter into an exchange-relation, so in order for the entire output to be consumed, the production and exchange relationships have to change:

Leyla's output   8 X_1 10 hours/day
Maria's output   2 X_2 10 hours/day
Total output $Q_{t+1} = q^L_{t+1} + q^M_{t+1} = 8\ X_1 + 2\ X_2 = 20$ hours/day   (6)

The "fair" exchange ratios according to the "labor-time spent" approach would be as follow:

Leyla's consumption C^L_{t+1} $= 4\ X_1 + 1\ X_2$ (7)
Maria's consumption C^M_{t+1} $= 4\ X_1 + 1\ X_2$ (8)
Total consumption $C^{L,M}_{t+1} = 8\ X_1 + 2\ X_2$ (9)

Both producers continue to work 10 hours a day, as in the initial case. But, due to the "new technology" developed by Leyla's creative mental labor, total output is increased:

$$Q_{t+1} > Q_t \qquad (10)$$

As is personal consumption:

$$C^L_{t+1} > C^L_t \qquad \text{and} \qquad C^M_{t+1} > C^M_t$$

However, this kind of "fair" exchange embodies an "unfair" feature. Although the increase in total consumption is entirely due to Leyla's contribution, the other consumer, Maria, who has so far contributed nothing, now benefits from the new situation just as much as Leyla. This kind of exchange relationship calculated in accordance with "labor-time employed" criterion does not seem "fair", at all.

Growth Function

In order to produce the capital-goods which embody the new technology introduced by Leyla's creative mental labor, the "given" supplies of nature have to be transformed into other physical products. By assumption, there were only two inputs of production; natural inputs and the labor-time employed which demonstrates that the supply of a means-of-production is a function of creative mental labor. Thus, the growth process (**g**) depends on the qualification level of Leyla's labor, **L**^{n}, or to be more precise, on her "creative" mental labor, **L**^{y}.

$$g = f(L^n) \quad \text{or} \quad g = f(L) \qquad (11)$$

Since, by assumption, there is no supply-demand imbalance, every output is consumed, but exchange-ratios vary.

The described "fair" exchange relationships in terms of the Marxist approach give rise to a serious deficiency. Maria, who makes no contribution to total output growth, benefits just as much as Leyla from the new situation. There is a serious error in this kind of growth process; Leyla's productivity is, in a sense, being penalized, while the "stationary" position of Maria is rewarded.

Barter-Exchange and Growth: 1-B

Assuming the presence of a new market for any additional output

In the model presented above, exchange relationships were based on the assumption that the community consisted of only two individuals. Now, we assume that additional items are sold to a new buyer in another market, while the exchange relationship between Leyla and Maria remains the same, as in the initial case.

Assume that Leyla exchanges the additional 4 units of X_1 with a third person and receives 3 units of X_3 in return. Under the new circumstances, the total output of the community, consisting of Leyla and Maria will increase, along with Leyla's output and consumption, while the position of Maria, as in the previous case, remains unchanged.

By assumption, there was no deficiency in "effective demand" for the additional output and both, **EE** and **TE** was at an optimum level:

      Leyla's output $8\ X_1$

      Maria's output $2\ X_2$

      Total output $Q_{t+1} = q^{L}{}_{t+1} + q^{M}{}_{t+1} = 8\ X_1 + 2\ X_2 = 20$ hours/day   (12)

      Maria's consumption $C^{M}{}_{t+1} = 2\ X_1 + 1\ X_2$ (13)

After trade with a third individual:

      Leyla's consumption $C^{L}{}_{t+1} = 2\ X_1 + 1\ X_2 + 3\ X_3$ (14)

      Total consumption $C^{L,M}{}_{t+1} = 4\ X_1 + 2\ X_2 + 3\ X_3$ (15)

      In other words;

         $Q_{t+1} > Q_t$

         $C^{L}{}_{t+1} > C^{L}{}_{t}$

But;

         $C^{M}{}_{t+1} = C^{M}{}_{t}$

Under the new circumstances, Leyla's consumption along with the total consumption of the community increases as a result of Leyla's contribution, while as for Maria, who made no contribution at all to the total output, her consumption remains "unchanged".

The "fair" exchange relationships in Marxist terminology are not valid anymore. The new exchange relationships seem to be more rational and fair, promoting the idea of further contributions which would benefit the consumer. This outcome has, actually, a much closer resemblance to actual market economic transactions.

Limits of Growth

By assumption, there was no deficiency in effective demand for the additional products. But, in fact, there is always a limit in the demand for the "available" products in the markets and the markets are bound to saturate sooner or later which implies the end of the growth process. Population growth may support the growth process to some extent, but not sufficiently, in the long term. There is always a limit to growth with "given" products. In other words, for the growth process to be continuous, the introduction of "**new products and production methods**" is imperative.

Barter-exchange and Growth -2

Technological productivity growth-2: New product and a new production method

In the simple model presented above, we studied how productivity growth affected exchange relationships, with a "given" product. Now, we shall assume that Leyla, by utilizing her "creative" mental labor, develops an "**entirely new product produced by a new production method**", denoted as X_4, in addition to the previous increase of the product X_1. Let us first analyze the production and exchange relationship between Leyla and Maria.

As in the previous models, we assume no lack of "effective demand", along with **EE** and **TE** being at optimum levels.

The new production function is:

$$Q = f(L_L^n; L_M^b) \qquad (16)$$

10 units of the new product (X_4) are produced, and the entire output is consumed in a domestic market consisting of Leyla and Maria. Since, by assumption the, markets are cleared and as a result of new technology the total output would increase. But as in the case of 1:a above, Maria will be the major beneficiary of output growth as a result of this barter exchange, although she has made no contribution at all.

Leyla's output $\quad$ $4 X_1 + 10 X_4$

Maria's output $\quad$ $2 X_2$

Total output $\quad$ $Q^T_{t+1} = 4 X_1 + 10 X_{4} + 2 X_2$ $\qquad$ (17)

The outcome of the "fair" barter-exchange based on the "labor-time employed" approach would be as follows:

Leyla's consumption $\quad$ $C^L_{t+1} = 2 X_1 + 5 X_4 \quad + 1 X_2$ $\qquad$ (18)

Maria's consumption $\quad$ $C^M_{t+1} = 2 X_1 + 5 X_4 \quad + 1 X_2$ $\qquad$ (19)

Total consumption $\quad$ $C^{L,M}_{t+1} = 4 X_1 + 10 X_4 + 2 X_2$ $\qquad$ (20)

"New Markets" for the "New Product"

Assuming a state of affairs where a new product is sold to a third party, along with a growth in productivity Leyla's consumption level will be improved, while Maria's will remain the same, as in the Model-1:b.

Let's analyze the following explanatory example: Assume that Leyla's "new" product, X_4, is used in trade with a third party and 5 units of X_4 is exchanged in return of 6 units of X_5. The new production and consumption relationships will be as follows:

Leyla's output $\quad$ $4 X_1 + 10 X_4$

Maria's output $\quad$ $2 X_2$

Total output $\quad$ $Q_{t+1} = 4 X_1 + 10 X_4 + 2 X_2$ $\qquad$ (21)

The outcome of barter-exchange:

Leyla's consumption $C^L_{t+1} = 2 X_1 + 1 X_2 + 5 X_4 + 6 X_5$ $\qquad$ (22)

Maria's consumption $\quad$ $C^M_{t+1} = 2 X_1 + 1 X_2$ $\qquad$ (23)

Total consumption $\quad$ $C^{L,M}_{t+1} = 4 X_1 + 2 X_2 \quad + 5X_4 + 6 X_5$ $\qquad$ (24)

To summarize;

$$Q_{t+1} > Q_t \qquad (25)$$
$$C^{L,M}_{t+1} > C^{L,M}_t \qquad (26)$$
$$C^L_{t+1} > C^L_t \qquad (27)$$

But;

$$C^M_{t+1} = C^M_t \qquad (28)$$

The interpretation of the above data is as follows: The reason for constant economic growth in the long term is the uninterrupted introduction of **entirely**

new products and production methods accompanied with an unsaturated demand for higher consumption.

The models studied above clearly show that the cause of all productivity increase is technological innovation which is the product of creative mental labor. Along with the increase in "**productive knowledge**", i.e. new technology, individual and total wealth increases. Since there seems to be no upper limit to the creativity of the human brain, there also seems to be no barriers, for now, for the long term growth of any economy.

Negative developments affecting environmental issues may reduce or exhaust completely the quantity of the necessary inputs of production, which, in turn, would bring an end to the growth process. But, it would not be irrational or illogical to expect that the necessary precautions would be introduced in time to prevent such a disaster. The decline in the global reserves of oil and coal should not emerge as a serious problem in the energy sector, because creative mental labor is capable of producing alternative energy sources. For instance, borax and hydrogen may be sources for future energy.

To summarize, one can easily claim that, assuming the supplies of nature are a "given", then the source of all added value that was accumulated in the past, is being accumulated at present and may be accumulated in the future is the result of human labor which constantly develops and uses these technologies in the production process.

Growth Process in the Real Economy

In the models presented so far, we studied production and distribution relationships under barter-exchange conditions where no money was used in these transactions, as used to occur in traditional economic analyses since the time of the Classical economists. Barter-exchange is no longer practiced in contemporary economies, but, nevertheless, it is used in the models above in order to explain the growth process. The analysis is believed to serve this specific purpose.

In this section of the study, an attempt will be made to reflect the actual output and growth relationship in a more realistic way. Some generalized and abstract statements, as well as some virtual economic relationships will certainly crop up but the intention is to make as realistic an analysis as possible in order to reflect real economic relationships.

The growth process can be studied under two sub-categories, as described in Gürak, 2006, Chapter-2:

Short term Growth: (given technology).
    1-a) **EE** and/or **TE** improvement.
    1-b) Production for new markets.
Long term Growth: (New technologies).
    2-a) A Given Product, but a New Production Method.
    2-b) A New Product and a New Production Method.

In the following sub-sections, the growth process will be studied with a "given" and "new" technology, respectively. An exchange model with money will be introduced later.

1- Growth without Technological Change: A "Given" Technology

1-a) EE and/or TE growth

The crucial assumption in the short term is production with a "given" technology, which implies an absence of technological innovation. Under such circumstances, if there is a lack in economic efficiency (**EE**) or technical efficiency (**TE**), in other words, if **EE** and/or **TE** are not at an optimum level, the output can be increased until it reaches the optimum level. But, once the optimum level is reached, growth ends, and the economy is in equilibrium.

At the optimum capacity utilization and efficiency level where there is no unemployment, output should be somewhere on the $Y_1 Y_2$ curve, as indicated in Figure-1. At point **X**, either **EE** or **TE** or both are not at maximum capacity. Output can be increased until it reaches the Y_1-Y_2 curve. But, then growth comes to an end.

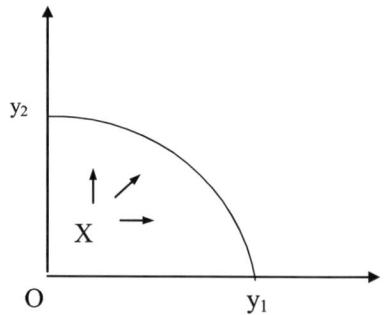

Figure-1   Short term growth with a "given" technology

138

Specific assumptions in this case are:

    1- A "Given" technology (T)

    2- The labor force is sufficiently qualified to use the "given" technology efficiently (L^k).

The production function is:

$$Q = f(L^k, X_i) \tag{29}$$

X_i denotes all inputs of production such as energy, raw-materials, marketing, the means of production, etc., excluding labor.

Short term growth is primarily a function of **EE** and/or **TE**.

$$g = f(EE, TE) \tag{30}$$

1-b) Extended production for new markets

Figure-1 can also be used to explain the finding of new markets for a "given" product. The equations (29) and (31) showing the output and growth functions, respectively, are also valid. In order to increase output, i.e., to grow, an increase in the inputs of production and of the labor employed would be sufficient, with the given technology.

If **EE** and **TE** are at optimum levels, then growth would be a function of **new investments**.

$$g = f(I) \tag{31}$$

This equation reflects the well-known orthodox economic approach, which dominated economic thinking until Solow's "**re-discovery**" of the vital role played by technology in the 1950s. With a "given" technology, investments take place and output increases; but only until the markets saturate. Then the growth process ends.

Investments are a function of the labor qualification level (L^n) and the expected rate of profit (r^E).

$$I = f(L^n, r^E) \tag{32}$$

2- Growth with "Technological Progress"

In some of the models above, the technology employed in production (T) and the qualification level of the laborer was assumed to be a "given". From now on, we shall focus on a case where the "creative" labor (L^y) enters the picture and paves the way for long term growth by introducing "new" technology. In other words, the long term growth takes place due to the technological innovations and the improvements in the qualifications of the individuals involved.

Assumptions:

    1- "Creative" mental labor (L^y) introduces new technologies (T^A).

    2- Education/training increases the qualification level of the laborers (L^n).

    3- Fair competition.

    4- **EE** and **TE** at an optimum level.

    5- No imbalance in supply-demand (**S=D**).

Now, let us study the impacts of technological innovations of a different nature.

2-a) A "Given" Product, and a "New" Production Method

Some "new" technologies are designed to produce a "given" product at a lower per unit cost. For instance, assume that a "given" product is produced at the unit cost of 10 TL by all producers. In a competitive environment, the producers would make every effort to produce the unit at less cost, in order to gain a cost advantage against their competitors. The producers who ignore this are bound to disappear from the market.

Assume that one of the competitors succeeds in reducing unit costs by employing a new production method. The reduction in unit cost may be due to one of the factors mentioned below:

    1- "Given" inputs, but output increases.

    2- Output increases faster than the increase in inputs.

    3- Output increases while inputs decrease.

    4- Inputs decrease while output remains the same.

The producer who achieves a cost advantage by using new technology would have three options.

    1- To reduce the sale-price.

    2- To increase the profits by maintaining the same price level.

    3- To follow a price-policy combining both the options mentioned above.

However, there is always an upper limit to growth with a "given" product though the producer may enjoy a competitive advantage due to the new technology, which reduces unit costs. Sooner or later, the markets are bound to saturate, the profit rate fall and growth rate will diminish and eventually stop.

The production function is:
$$Q = f(L^y, L^k, L^b, X_i, T^A) \qquad (33)$$
Or simply:
$$Q = f(L^n, X_i, T^A) \qquad (34)$$

L^n, denotes the labor-force with a sufficient level of qualification to employ the new technology efficiently. As we know, technological innovations are introduced by L^y in order to increase the productivity of the labor force and are embodied in the physical products. Therefore, there is only one "productive" factor in the production function; $\mathbf{L^n}$ or just $\mathbf{L}$.

Technological innovations (T^A), are internalized (embodied) in the physical products employed in the production process ($\mathbf{X_i}$) and are a function of "creative" mental labor ($\mathbf{L^y}$), the technological development level ($\mathbf{T}$) and R&D investments ($\mathbf{I^{R\text{-}D}}$):

$$T^A = f(L^y, T, I^{R\text{-}D}) \tag{35}$$

Assuming that there is no depreciation of the "means of production", the growth function would be:

$$\mathbf{g = f(I^T)} \tag{36}$$
$$I^T = f(L^n, r^E) \tag{37}$$

$\mathbf{I^T}$ denotes "new" investment due to technological innovation, $\mathbf{L^n}$ the laborers required to use the new technology, and $\mathbf{r^E}$ the expected profit rate. The laborer requires further qualification for the efficient employment of the new technologies. Therefore, for the growth process to be complete, the qualification level of the laborer has to be increased in addition to a new level of investment.

When new investments are introduced in order to replace the depreciated "means of production" ($\mathbf{I^d}$) in addition to the "new investment" due to technological progress ($\mathbf{I^T}$), the growth function would be:

$$g = f(I^T, I^d) \tag{38}$$

As displayed in Figure: 2, demand would increase somewhat, assuming that there is a price fall as a result of the cost-reducing technological innovation. But, nevertheless, the growth process will eventually come to an end, when the markets saturate.

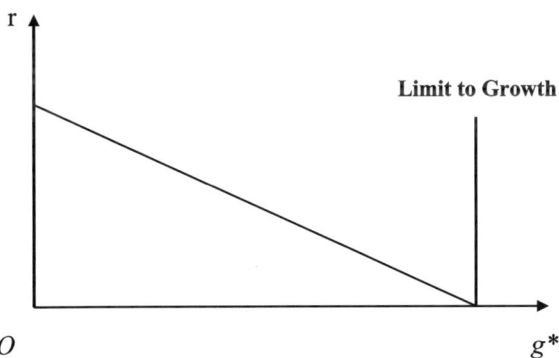

Figure: 2 Limits to growth and profit rate (given technology)

2-b) New product, new production method

Long term growth is a dynamic process without a foreseeable limit, though it may, from time to time, experience some disruption. There seems to be no upper limit to the creative capabilities of mental labor in creating new products and/or production methods; nor does there seem to be an upper limit in the demand for new products. Under such circumstances, there would be no grounds to expect markets to saturate, nor for the growth process to end. The profit-rate would follow a trajectory with peaks and troughs, in accordance with technological innovation. Now, we shall study how the introduction of new products influence the growth process, assuming **EE** and **TE** are at an optimum level. As before, **S=D** by assumption.

Production function:

$$Q = f(L^n, X_i, T^A) \tag{39}$$

Assuming no depreciation of the means of production, the growth function would be:

$$g = f(I^T) \tag{40}$$
$$I^T = f(L^n, r^E) \tag{41}$$

As displayed in Figure: 3, the profit-rate would never fall to zero level given the constant supply of technological innovations. However, it is expected to fluctuate. To put it another way, in the initial phase of a new technology, the owner will enjoy "monopoly" privileges in the market and is very likely to obtain a higher profit-rate above the average for the market. In time, the competitors are assumed to introduce similar technologies and products, which would cause the monopoly profits to decline. But, a wave of new technologies would again facilitate monopoly profits, and the process would continue in this manner. Therefore, the profit-rate is never expected to decline to a zero level, but to fluctuate.

When new investments are introduced in order to replace the depreciation of the means of production (I^d) and in addition to new investments due to technological changes (I^T) the growth function would be as in eq.42.

$$g = f(I^T, I^d) \tag{42}$$

Profit rate

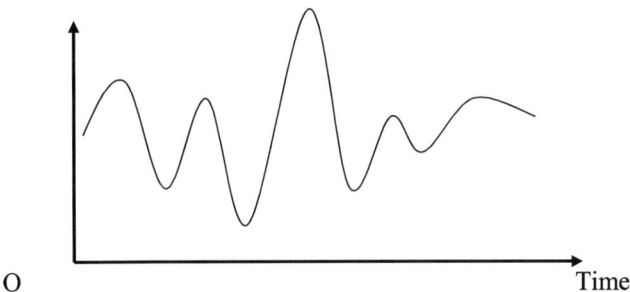

Figure: 3 Probable fluctuations in the rate of profit.

To summarize; as long as demand for new products does not cease, new products will continue to be developed by mental labor, thereby increasing output and consumption, thus paving the way for a constant growth process in the long term, with the inevitable fluctuations, of course.

Growth: Reconsidered - Both in the Short- and Long Term

In the models above, we first studied the short term growth process in the absence of technological innovations, and then long term growth with technological innovations. Naturally, in real economies the individuals do not make decisions in line with the models. Producers normally begin a production process in regard to their long term expectations, though they often face some unexpected incidents in the short term and might have to make some critical decisions. In fact, the long term outcome consists of short term processes. The question is: Is it possible to a have a production and growth function covering both periods, i.e., the short and long term.

Regardless of the time aspect, whether short or long term, the inputs of production, or, in line with the orthodox terminology, "the production factors", are always "the same".

$$Q = f(L^n, X_i) \tag{43}$$

However, the case for the growth function is somewhat different. Because, we now have to take into account the growth process in both the short and long term.

$$g = f(EE, TE, I) \tag{44}$$
$$I = f(I^T, I^d, L, r^E) \tag{45}$$

I, denotes both, new investments due to technological progress, $\mathbf{I^T}$, and new investments to replace the depreciated means of production, $\mathbf{I^d}$, while **L** denotes the qualified labor force required to efficiently realize production, and $\mathbf{r^E}$ denotes the expected profit rate.

The Role Played By Demand in the Growth Process

The assumption that supply declines as the markets saturate, which in turn causes demand to fall is a crucial one. Because fluctuations in demand and supply are not necessarily proportional. Sometimes the quantity demanded precedes supply, and sometimes vice versa. There are many factors influencing the demand schedule such as:

    1- The price of a product.

    2- The purchasing power of the consumers (income level).

    3- Tastes and preferences.

    4- The prices of competing products.

    5- Public expenditure.

The role played by demand in the growth process is naturally significant. However, in the absence of technological innovation which introduces "new products", the demand for goods and services cannot alone secure long term growth. Assume that a "given product" is produced with a "given technology". In time, as the markets gradually saturate, demand will gradually decline and eventually there will only be some production for the replacement of the depreciated goods. In other words, there will be "equilibrium" in markets. Assume that only cost-reducing technological innovations take place and are used in production. The consequence will be the same as in the previous case, i.e., as the markets saturate, the growth process will gradually decline until equilibrium, given the purchasing power. Therefore, in both cases, in the absence of technological innovations that introduce "new products", demand will not be a sufficient condition to sustain long term growth.

Since the emphasis in this study has been one of revealing the source of long term growth. A thorough analysis of the impact of demand on the growth process will not be entered into at this time.

Conclusions

According to the findings of this section, laborer appears as the only production factor capable of adding value, assuming the supplies from nature (the basic inputs of production) as both stable and a "given". Labor transforms nature's supplies into useful products to be sold in the markets. All "means-of-production" are transformed natural inputs aimed at increasing labor-productivity. Thus, labor, or rather the qualification level of the labor is the source of all the added value of a product as well as all new technology required for any growth to take place.

As to the qualification level of the labor force; it is proportional to the technological development level of the country in which they live and also to the quality of the education acquired. The increase in the number of technological innovations and in the level of general welfare of a nation is only achieved by the contribution of the creative mental abilities of the labor force.

In any theory in regard to value-creation and growth, it is essential to distinguish between the qualified and non-qualified laborers with regard to their contributions to production. Laborers endowed with qualifications (or similarly with human capital) may be **"technology-using"** laborers, which is important for the efficient use of any given technology. If the qualification level of laborer falls short of expectations, the outcome would inevitably lead to inefficiency and a waste of resources. Therefore, it is not only necessary to be endowed with a certain level of qualification but it is also imperative to have the "right" endowment in order to avoid inefficiency in the supply of products. Alternatively, given the technology, there is a direct relationship between the capabilities of the laborer and the level of advancement of a country.

Above reference was made to **"technology-using"** laborers. In the long term where technological progress takes place, it is imperative to have access to **"technology-creating"** laborers in order to increase the total and individual wealth in an economy. As we know, technological progress is the source of all long term growth while **creative mental labor** is the source of all technological progress.

Reference

Gürak, H.   2006   Ekonomik büyüme ve küresel ekonomi
                         Ekin Kitabevi, Bursa.

5- TECHNOLOGICAL PROGRESS & GROWTH

"Technological Productivity" and Long-term Growth

Introduction

The relations between growth, technological progress, productivity and labor(-er) were analyzed in other works (Gürak, 2000-a, 2000-b, 2004-a, 2004-b), in order to determine which inputs of production such as laborer, raw-materials, capital-goods, intermediaries, were regarded as **"factors employed in production"**. There were only **"two productive factors"**; **"nature"** and **"labor(-er)"**, and the latter was shown to be the only productive factor capable of producing an **"exchange-value"** or an **"added value"**. In this analysis, the same premises will be valid. Thus, given the necessary supplies of nature (raw materials) for output, the analysis will focus on the relationship between "technological progress" and "economic growth".

Human labor can be analyzed into two categories;

1- mental; and
2- physical.

Physical labor contributes only to a limited extent in value creation, while mental labor together with its creative ability, is the actual source in the creation of value and makes an indispensable contribution to a constantly growing productivity and global welfare. Mental labor makes its "long-term" contribution by introducing **"new productive knowledge"** i.e., technology. This new productive knowledge which is utilized by mental and physical labor helps to transform natural supplies to produce either a "means-of-production" which in turn leads to an increase in labor productivity or is used to supply intermediary or consumption items.

If mental labor had made no contribution to the ever growing pool of knowledge by incessantly introducing "new productive knowledge" (e.g., new technologies), then neither the increase in the total supply of useful products, nor the living standards of humans would not be any different than those existing hundreds of years ago. This, in turn means that humans would now be living at a basic subsistent level comprising simply of finding nourishment, ensuring protection and finding shelter. In fact, until about 300 years ago human living standards were not much different to that existed in medieval times (see Figure-1).

"The estimates indicate that per capita GDP in 1700 was rather similar around the world and there were minor differences between USA, China and India. There was nearly no change in the per capita GDP between 1700 and 1820." (Vasquez, 2003, p.90).

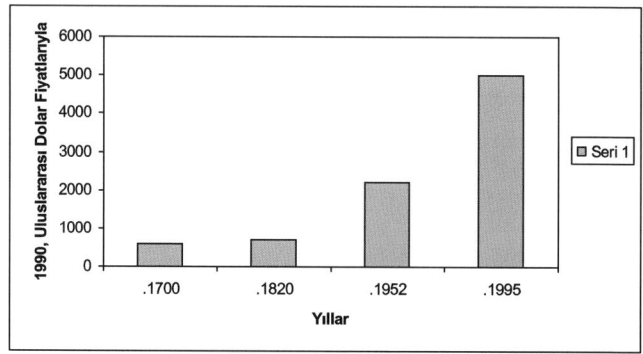

Source: A. Maddison, Monitoring the World Economy: 1820-1992;

Figure: 1   World per capita GDP

Since the 1700s, productivity growth and the standards of living began to diverge between countries.

"Per capita income doubled in Europe and the USA, in comparison to China, Japan or Russia... Then, the economic expansion in the 19th. Century multiplied the living standards by three times in Europe and four times in the USA." (Vasquez; 2003; p.90).

Table:1 shows very clearly how 25 extra-ordinary events in the USA affected the total welfare and qualitative structure of society in the 20th century.

Due to the "new technologies" originating from mental labor, both individual and total welfare increases continuously. In spite of various environmental problems Mother Nature continues to supply the necessary inputs of production, while laborers transforms them into useful products with the assistance of technology. This process will, apparently, continue as long as the creative abilities of mental labor continue.

Table-1 25 Extra-ordinary events in the USA in 20. Century

Events	1900-20[a]	1995-98[b]
Life expectancy (years)	47	77
Infant mortality (per 1000 birth)	100	7
Deaths from contagious diseases (per 100,000 persons)	700	50
Heart diseases (per 100,000 persons-age-adjusted mortality)	307 (1950)	126
Per capita GDP (1998, US $)	4,800 $	31,500 $
Wages in manufacturing sector (1998, US $)	3.40 $	12.5O $
Assets per household (1998, trillion US $)	6 $ (1945)	41 $
Poverty rate (as % of US households)	40	13
Weekly work-period (hours)	50	35
Agricultural labor force (as % of total labor force)	35	2.5
TV-set ownership (as % of US households)	0	98
House ownership (as % of US households)	46	66
Electrification (as % of US households)	8	99
Phone calls (per capita annual calls)	40	2,300
Transport vehicles (as % of US households)	1	91
Patents granted	25,000	150,000
High school graduation (as % of adults)	22	88
Deaths from accidents (per 100,000 persons)	88	34
Wheat price (worked hours per kilogram)	4.1	0.2
Faculty degrees granted to women (as % of total)	34	55
Income of blacks (annually, per capita, in 1997, US $)	1,200 $	12,400 $
Resident population in the USA (millions)	76	265
Air pollution (microgram lead per 100 m³ air)	135 (1977)	4
Computer speed (million directives per second)	0. 02 (1976)	700
Computer ownership (as % of US households)	1 (1980)	44

a: Values refer to eldest data available.

b: Values refer to newest data available.

Source: Vasquez, I.O (2003), Kapitalizm ve Küresel Refah, p.62, Table 4.1.

The Purpose of the Study

The purpose of this work is to analyze the interrelation between the growth in the value-added output (VA) and the producers' technological innovations.

Products supplied for personal consumption are beyond the scope of this work. For instance, a value created by cooking at home for personal needs or repairing the broken leg of table is not considered as VA-creation. Eating in a restaurant or having the broken table leg fixed by professionals, on the other hand, is considered to be within the scope of this analysis. According to this definition, as the produced **VA** increases, both individual and total riches increase. In other words, the attempts of firms to increase productivity in order to increase their profits lead to economic growth, cet. par.

One of the subjects to be discussed within the framework of the long-term increase in productivity (growth) is how the added value (**VA**), the price of a product (**p**), the rate of profit (**r**) and functional income distribution is affected by **technological innovations**, which in their turn are products of "**mental labor**".

The term products refer to "physical" goods (commodities), as is traditionally accepted in economic models. Yet, in modern societies, the service sector constitutes a larger share of the GDP than the industrial or agricultural sectors and this share is increasing. As service sector production relationships display different features, the impact of technological innovation on price-formation, **VA** creation, profit-rate and income distribution will be considered separately.

Assumptions:

- a "fair" competitive environment (but it does not have to be perfect);
- no shortage of qualified labor, e.g. human capital;
- no shortage of natural or financial resources;
- a contemporary institutional and cultural infrastructure;
- no market interventions;
- an optimum level of efficiency (micro-productivity);
- no inflation;
- gross profits contain interest payments;
- total costs contain depreciation costs;
- the supply-demand relationship is stable (no shortage or insufficiency of demand).

We shall start with some basic definitions and explanations in regard to the concept of "productivity" and its measurement. Then, the relationship between mental labor and productivity will be studied, which is considered crucial to the understanding of the growth process. Afterwards, key concepts such as "efficiency" and "technological productivity" will be discussed. As technological innovations are the cause of long term economic growth, the concept of "technological productivity" will be studied thoroughly.

A concise note on demand

The role played by demand in the growth process is naturally very significant. However, in the absence of a technological innovation which introduces "new products", the demand for goods and/or services cannot alone secure long term growth. Assume that a "given product" is produced by a "given technology". In time, as the markets gradually saturate, demand will gradually decline and eventually there would only be production for the replacement of depreciated goods. In other words, there will be "equilibrium" in the markets. If one assumes that only cost-reducing technological innovations take place and are used in production then the consequences will be the same as in the previous case, i.e., given the purchasing power, as the markets saturate, the growth process will gradually decline until it reaches a state of supply-demand equilibrium. Therefore, in both cases, in the absence of technological innovation which leads to the introduction of "new products", demand will not be a sufficient condition to sustain long term growth.

In this work, the emphasis will be on revealing "the source" of long term growth. Therefore, a thorough analysis of the impact of demand on the growth process will be ignored.

Productivity, Productivity Growth & the "Measurement" Problem

Productivity (V)

Productivity is **a** static concept and refers to the relationship between the "inputs" and "outputs" of production. Therefore, productivity can be described as the "**capability to produce**" or, within the context of economics, as the "**capability to produce VA**" especially in a country or an inter-country analysis. Enterprises would be more interested in making and maximizing profits rather than focusing on added value. Therefore, unless otherwise stated, we shall be referring to "the added value" concept in our analysis in the following sections.

The analysis of "productivity" can be carried out from two different aspects;
1- Quantities;
2- Values.

Quantitative Analysis of "Productivity"

Any quantitative analysis of productivity contains some inconveniences and difficulties. For instance, assuming only one type of output, e.g., a homogenous product, there would be no serious difficulties in measuring the productivity per employee, or "partial" productivity (KV) in terms of any "given input". When heterogeneous outputs or more than one input are involved, measurement becomes cumbersome, even impossible. Assume that 110 tires are produced in a plant and the inputs of production are 10 workers and 20 Kg of rubber, 100 Kw energy and two machines. Partial productivity (**KV**) with reference to the employee can be shown as "output per employee".

KV = Output per Employees = 110 / 10 = **11 Tires Per employee** (1)
Or, alternatively:
KV = 110 /20 Kg Rubber = **5.5 Tires / 1 Kg Rubber** (2)
Equation (1) shows that one employee produces 11 pieces of tires, and equation (2) shows that 5.5 pieces of tires are produced with 1 Kg rubber.

But, what if the inputs are more than one? Would it still be feasible to measure the total or partial productivity? Equations (3) and (4) give the answer.

V = 110 pieces tires /10 employees+ 10 Kg rubber+2 machines= **???** (3)
KV = 110 pieces tires/ 10 employees + 10 Kg rubber = **???** (4)

It would not be absurd to claim that it is not possible to measure the total input productivity (V) in terms of quantities.

V = Output / Employees + All non-labor inputs = **???** (5)
There is more to it.

Let us re-consider the output per employee analysis in terms of quantity and take a look at the output of two competitive firms in the auto industry. Assume that one worker at the Mercedes plant produces 5 cars while at the Ford plant one worker produces 7 cars a day. Is the worker at the Ford automobile plant more productive than the worker at the Mercedes plant?

Certainly, not. It would be irrational and misleading to make such a comparison. For a sound comparative analysis the output of both firms have to be the same, e.g., homogeneous, and employ the same technology. This is a case encountered in the neoclassical economic analysis, only.

In short, a quantitative analysis of productivity with two or more inputs of production is not of much value, while partial analysis with regard to one input only can give us some useful insights.

Value analysis of "productivity"

Measuring productivity in terms of value is not free of problems, either. But, it seems to be less cumbersome, more reliable and a less complex method to use. In measuring productivity, a specified value criterion, say added value (VA = wage + profit) can be used. Using the VA criterion, the productivity analysis can be divided into 5 broad sub-categories:

V = VA / TM = Wage + profit / Total production cost
KV = VA / L = Wage + profit / Per employee
KV = VA / t = Wage + profit / Unit time employed
KV = VA / W = Wage + profit / Total wage bill
KV = VA / OC = Wage + profit / Non-wage input costs

TM, denotes total production costs; L, employees; t, unit time employed; W, ($L*w$) total wage bill; and OC, the non-wage input costs.

"Productivity Growth" (Economic Growth)

Productivity growth is a dynamic concept indicating an increase in the products supplied, leading to economic growth. It is a vital fact of economic life, because, not only does it improve the competitive strength of the producer, but also increases the total added value, and in turn the total economic welfare.

The study of the causes of productivity growth is important. But, since a thorough study has already been done in another article (Gürak,2004-b), only a brief reference will be made here.

Productivity growth in terms of value can be analyzed in two sub-categories: added value (**VA**) and quantities (**Q**).

    1- Quantitative growth: or

    2- Value growth.

A quantitative growth indicates a variation in the relationship between the quantities used in production and the quantities supplied. The remarks made on the "quantitative productivity" analysis are also valid for a "quantitative productivity growth" analysis. In other words, the measurement of productivity growth in regard to two or more inputs would produce neither healthy nor reliable results. A measurement in regard to one input only, especially in regard to employees or the unit time employed would be possible, but the outcome is bound to be dubious, especially in regard to "new" products and "production methods".

Measuring productivity growth based on monetary value, or to be more specific, on the added value criterion for national economies seems to be much less complicated and displays less serious shortcomings compared to the quantitative productivity growth analysis. Therefore, from now on, productivity growth will be considered only in terms of the added value produced (**VA**) in a national economy, unless otherwise stated. Since laborers are regarded as the only production factor capable of adding value, it seems rational to measure productivity in terms of the value added (**VA/L**) or in terms of total income (**TR/L**) in regard to the employed worker, or in regard to the wage cost (**VA/LWC or TR/LWC**), respectively. In this work the latter method (**VA/LWC**) will be preferred in the analysis in regard to national productivity growth.

Is the "value" criterion a perfect choice?

A "productivity growth" analysis in terms of value is, certainly, not free of controversy, but it is more likely to produce more reliable results compared to a quantitative analysis. For instance, assume that the national currency of a country for some reason changes in value, cet. par. In terms of the value criterion, it would appear that the productivity in this particular country has changed, although there does not need to be any variations in the production relations within the country. This situation reflects one of the shortcomings of the value criterion. But it seems, nevertheless, less cumbersome than the quantitative analysis.

In this work, it is assumed that producers employ their resources at an optimum efficiency level in an economy free of inflation or devaluation risks. In other words, there is optimum micro-economic efficiency in production. The crucial distinction is; the system is characterized by "**technological productivity growth**" caused by "**new technologies**" or, synonymously, by "**technological innovations**". And we shall study how this process affects **the "production costs"**, the "**price**", "**profitability**", as well as "**income distribution**".

The Relationships between Mental Labor, Technological Innovation & Growth

The premise that human creative mental labor is the incessant source of value-creation as well as being the only productive input capable of adding value, has already been discussed elsewhere (Gürak, 2000-a; and 2004-b). We will now focus on the discussion of other key concepts such as education, training, skill, experience, a qualified labor force and creative mental labor.

The laborer employed in production can be divided into two categories:
    1- Laborer with a level of quality (human capital); and
    2- Laborer without quality (no human capital).

Laborer with a level of quality (human capital) as a concept refers to the set of knowledge acquired through both formal and informal education and training, personal skills and abilities as well as the experience acquired by the laborer. Thus, a **qualified laborer** or, alternatively, a **laborer with human capital** is the person who possesses some or all of these features to certain extent; and the **qualified labor force** or, alternatively, the **labor-force with human capital** refers to the sum of the laborers with such endowments. In other words, **qualified labor (work)** is the **"service" supplied by qualified laborers**.

The existing state of general, scientific and technological knowledge is determined by knowledge accumulated in the past and any present contributions to this pool of knowledge. To measure the extent of this present knowledge and the qualifications of a society or labor-force, various methods can be used. For example, the total or average time spent on education or training could be a yardstick, but not a perfect one. As is well known it is virtually impossible to correctly measure the quality of education and/or training, or the teaching and learning abilities of teachers or students, respectively. Nor could a proper comparison be feasible.

In addition, there is no precise measurement with which to accurately compare the "experiences" of different individuals. The degree and quality of experience cannot be determined simply by the level of his or her education and/or training but must also consider their personal abilities and years of work. Thus, measuring a concept such as "experience" analytically using mathematical or statistical tools is unlikely to produce sound results.

Technology was defined as "the knowledge created by mental labor used in order to modify and to control the environment in which we live (Gürak; 2004-b). Accordingly, the qualifications possessed by the laborers (i.e., education and/or training plus skills, plus experience) play a vital role both in the creation of new technologies and the efficient application of those technologies presently in use.

The critical question in this relationship is; what is the role played by "unqualified" laborers?

With reference to the definition of **qualified laborer**, the term **unqualified laborer** should refer to the "uneducated/untrained, unskilled and inexperienced" laborer, which is practically and scientifically out of the question. An "entirely unqualified" laborer can only exist in artificial economic models, not in reality. In contemporary societies, every individual possesses some level of qualification acquired through formal or informal education and/or training, skills and experience. A rational and realistic theoretical economic model and analysis should mention the differences in the level of qualification. In reality, there exists no laborer without some degree of qualification. Some laborers may not have the proper qualifications for a specific job, but that is an entirely separate matter. A more correct definition could be stated as "**qualified or less qualified**" or alternatively "**qualified or insufficiently qualified.**" Therefore, the classification of laborers into "qualified and unqualified" or as is found in the orthodox doctrines as labor (**L**) and human capital (**H**) is artificial and irrational and prevents any proper economic analyses.

The technological and general economic development level of a country is closely related to the total qualifications of its labor-force. In other words, the higher the total qualifications, the higher the country's overall development level is expected to be. Not only are "new technology-producing creative minds" essential but also "efficient technology-using minds".

However, "technology-using" qualifications alone are not sufficient to secure **long term** economic growth. For instance, assume that there is a nation that possesses a sufficient number of qualified laborers in every branch of industry and employs them efficiently in a production process of "given" goods using the "given" technologies. However, some day, the markets for the "given" products will saturate. In order to sustain growth in the long term, **new technologies** producing **new products and/or production methods will have to be introduced.** And to produce new technologies, **a qualified labor force with creative abilities** will be required. Thus, **creative mental labor** comes the fore as the most important source of long term economic growth.

The Transfer of Technology

The Less Developing Countries (LDCs) do not necessarily need a supply of technology-producing creative mental labor in order to realize economic growth. In other words, the presence of creative minds producing new technologies is not an imperative, at least not for short and medium term growth. A large number of **"new technologies for LDCs"** already exists in the Developed Countries (DCs). What the LDCs need more is a qualified labor force capable of using existing technologies efficiently, which require the proper global markets for technology transfer. "Re-inventing" already invented technology which is presently available in the DCs is not only highly costly and risky, but also irrational. Therefore, it would be more rational, if, rather than producing entirely new technologies the LDCs devoted their scarce resources to improving the qualification level of their labor force aiming for increasing efficiency. In specific cases, the choice of a rational policy would vary from country to country depending on the specific conditions in the particular country. Unfortunately global technology markets are far from perfect and are designed primarily to protect and promote the interest of the DC firms, rather than promoting growth in global competition and welfare. Restrictive clauses and especially the **"transfer pricing"** mechanism, which accompanies the technology transfer process contains many features which work against the interests of the LDCs and against the promotion of global competition. In order to increase the benefits for the LDCs from a technology transfer, the prevailing market imperfections have to be corrected and a new system has to be introduced which promotes global competitiveness, which in turn would make the world an economically a better place to live (see Gürak, 1990).

The Limit of the Growth

Assume that a nation with some "given" technologies but which produces no technological innovation and where production is at an optimum productivity level. That is to say all human, physical and financial resources are employed at a maximum efficiency level. In that nation, "output" as well as "welfare" would be at the highest achievable level. The growth rate of the economy would depend only on the population growth rate. Yet, in reality, economic growth is a

156

dynamic and continuous process independent of the population growth rate. It is a function of "productive knowledge".

The most significant distinction which separates human beings from other species is the "production of knowledge". Since the Industrial Revolution in Britain, the supply and accumulation of knowledge has increased rapidly. And with the impressive developments in such sectors as electronics, computers and semi-conductors, not only the supply but also the dissemination and exchange of information and knowledge have begun to increase rapidly. The growing pool of productive knowledge (new technologies) increases the demand for laborers with higher qualifications in order to employ the new technologies efficiently. Nowadays it is imperative to have access to a highly qualified labor force not only to maintain high standards of living in the DCs, but also for the LDCs to catch-up or close the gap with the DCs. Learning new things has now become a **"permanent"** process for both societies and firms.

"Productive knowledge" is not the only factor influencing growth. Other major factors are:

1- Conditions for "fair" global and domestic competition.
2- An appropriate institutional, political and cultural infrastructure.
3- A certain level of technological development.
4- Financial resources.
5- Natural resources.

Excluding nature and laborer, the rest is not capable of producing any value by itself and cannot make any direct contribution to the growth process, except indirectly. Laborer is the only input of production capable of "adding value" to the available products. Assume there is a total destruction of all the man-made physical products due to some cataclysm, if natural resources remain then everything can be reproduced by knowledgeable laborers.

A significant proportion of man-made physical items were destroyed both in Japan and Germany during WW II. But both countries recovered rapidly and have taken their place among the most developed and industrialized nations of the world. Both had access to "technologies" and a "qualified labor force" required for the successful reconstruction of everything that was destroyed. On the other hand some countries like Turkey did not have access to the necessary technologies or a qualified labor force, did not participate in the war and did not suffer any physical destruction, but are still struggling to catch-up and close the development gap. South Korea, which had a lower per capita income than Turkey in the 1950s and 1960s, has nowadays not only closed the development gap but also surpassed Turkey by a good margin. In 2003, the real GDP in Turkey was $ 6,700 while in S. Korea it was about $ 15,700 (Derviş, 2011). The explanation for this is the difference in the qualification level of their respective la-bor-forces.

Efficiency and Technological-Productivity Growth

Because of competition, producers while striving for higher profits, constantly seek for new opportunities to reduce costs and/or to introduce new products. These competitive attempts to increase productivity also contribute to the growth of welfare and shall be studied below under two subcategories "**micro**" and "**macro**" productivity growth in terms of value.

Growth-1: Efficiency (Micro-Productivity) Increase

Productivity increase does not only take place due to the introduction of new technologies. Improvements in the efficient use of the available resources with a given technology can also lead to an increase in productivity. Therefore, it is imperative to draw a distinction in the nature of these productivity increases. Increases in productivity with "given" technologies shall be referred to as "micro productivity", or synonymously, as "**efficiency growth**" and is only appli-cable to **short-term growth** (see Table:2).

The principle determinants of efficiency growth are given in Table:3. Alt-hough no new technologies are introduced, efficiency growth contributes to in-crease in the produced added value (**VA**), the rate of profit (**r**) and the share of profit in **VA**, while causing the share of wage (**w**) in **VA** to fall, cet. par.

Table: 2 The causes of growth in efficiency (micro-productivity) and technological (macro) productivity

Macro (technological) productivity growth
  New method of production; or
  New product and production method
	Long-term	**New** technology

Micro productivity (efficiency) growth
  1- Human resource efficiency
  2- Financial efficiency
  2- Technical efficiency
	Short-term	"**Given**" technology

Table: 3 Relationships between efficiency (micro-productivity) and growth, added value (VA) and profitability (r)

New tech-nology	Cause of efficiency (micro-productivity) growth	VA/K	VA/L	r	π/VA	w/VA
No	Restructuring production	↑	↑	↑	↑	↓
No	Increasing capacity utilization	↑	↑	↑	↑	↓
No	Multiple shift-work	↑	↑	↑	↑	↓
No	Reallocating the re-sources	↑	↑	↑	↑	↓
No	Improved general educa-tion	↑	↑	↑	↑	↓
No	Learning on-the-job & experience	↑	↑	↑	↑	↓
No	Improved safety & sanitary	↑	↑	↑	↑	↓
No	Democracy at plant-site	↑	↑	↑	↑	↓

The Limits of "Efficiency (Micro-Productivity) Growth"

Assume that a firm produces goods (**X**), and all resources, human, physical and financial, are employed at an optimum level, that is maximum technical efficiency prevails and profits are maximized while costs are minimized. What are the options for firms if they desire to increase their profits?

1- **Horizontal growth:** One of the options is to find **new markets** for their products. In order to meet the potential demand for "given" goods from the new markets, the producer may have to make "expansive" investments". As long as demand grows, the income and thus total profits of a firm will continue to increase, cet. par. However, given the product, there is always an upper limit for demand. As the market for a given product begins to saturate, the strength of demand will begin to decline and eventually halt. After that point, the output could only aim to meet a demand caused by depreciation and population growth, given the existing purchasing-power. The impact of population growth on output growth can only be marginal. Thus, the growth process without "new products" is sooner or later bound to come to an end, as the more pessimistic theories predict.

2- Wage cut: Another option for a firm to increase its profits is to reduce the wages paid to employees, cet. par. Thus, the share of wages in the added value accrued will drop and the share of profits increases, while the total added value remains unchanged. This option may prove beneficial for the wage-reducing firm, but if all the firms introduced wage cuts at the same time it would likely produce opposite results for the economy. It would reduce the total demand and the total added value produced, implying a negative growth. Therefore, what is beneficial for one firm is not necessarily beneficial for the other firms or the economy.

3- Reducing the cost of production: Given the technology and an optimum efficiency level in production, the only possibility to reduce costs seems to be paying less for the purchased inputs of production. But, the suppliers of these inputs would naturally be reluctant to sell their inputs for a lower price, which would lead to their profits being reduced, cet. par.

According to the cases discussed above, in the absence of technological progress, there is an inevitable limit to growth with "given" products. Assuming less than optimum efficiency in the production process, there is scope to increase the supply of added value i.e. growth. But, as the markets approach saturation point, the strength of demand is bound to decline and eventually stop. Therefore, it is imperative to introduce technological innovations to sustain long term economic growth.

Growth-2: Technological Productivity Growth

Compared to efficiency (micro-productivity) growth, technological growth contains a distinguishing and crucial feature, i.e., **"new technologies"** or, synonymously, **"technological innovations**. It implies that the economy is now subject to changes in terms of "new" products and/or to the unit cost reducing production methods of the "given" products. Technological productivity growth, which is the driving force of long term economic growth, can be divided into two groups (see Table:4):

Table: 4 Technological (macro) productivity growth

1- Given product but new production method 2- New product/new production method	Long-run	New technology

A-) A "Given" Product but a "New" Production Process

Given the product, the most rational and effective behavior of firms in order to improve their competitive strength and profits is to employ a new unit cost reducing production method. To serve this purpose, the new technology should facilitate production with one of the following features:

1- The same amount of **VA**, but with less inputs or labor.
2- The same amount and value of inputs, but more **VA.**
3- The growth rate of the **VA** is greater than the growth rate of the value of the inputs.
4- The rate of decline in the value of the inputs is greater than the rate of decline in the **VA.**

A related problem may occur with the strength of the demand. As the product is assumed to be given, demand is expected to decline in relation to the saturation of the market, which would cause the profit rate to decline, which, in turn, would cause a decline in the investment rate.

B-) A New product and/or a New Production Method

The distinguishing feature of this kind of macro (technological) productivity is the introduction of "new products" usually accompanied by "new production methods". Long term sustainable economic growth can only occur through an increase in technological productivity. Long term economic growth rate is normally subject to fluctuations, and sometimes even crises occur. However neither does the growth process cease, nor does the profit rate tend to fall to zero in the long term. Often, the shortage or exhaustion of certain natural resources becomes the major subject of economic debates. But, nevertheless, the economic growth process continues. Moreover, in contrast to approximately the 3,000 hours worked annually 50 or 60 years ago, an employee works about 1,500 to 2,000 hours nowadays and produces much better quality and a greater variety of products. The only reason for these developments is the incessant **technological productivity growth** which supplies new technologies, or, to be more specific, the incessant supply of **new products and/or new production methods**.

New products require new investments, which in turn naturally generates new demand. As a result, the initial profit rate tends to be higher than the average rate prevailing in the markets with a given product or technology. This is a natural process. Otherwise, the firms would have no incentive to take risks to utilize the new technologies. As a result of new technology, not only does the profit rate rise, but the options for the consumer increases.

To summarize, the reason that the long term profit rate does not to fall to zero and economic growth is sustained is due to technological productivity growth. New inventions and the application of new technology is an irreversible and uninterruptible process of economic life in progressive societies. Unceasing technological productivity growth serves not only the interests of the producer but also those of the individual. As a result of technological productivity growth, employees work less hours, produce more and better quality products. Additionally, the real income per capita increases, leading to more consumption and an improvement in living standards.   In spite of some setbacks and objections, people in general, welcome such developments.

Why Do Firms Need Technological Productivity Growth?

Three major reasons can be suggested to explain why competitive profit-maximizing firms search for "new products" or "new production methods" in order to enable their technological productivity to grow. These are:
　　1- Competitive advantage. To be ahead of the competition.
　　2- Monopoly advantage. To maximize long term profits.
　　3- Defensive strategy. Not to lag behind the competition.
　　The main goal of every commercial firm in a competitive environment is to first survive the competition and then to maximize the profit rate and long term profits. In the short term, a commercial firm may increase its profits by reducing some of the production costs through some non-technological measures. But, in the long term, the only way it can keep profits from falling to zero is to introduce new products and new production methods that means increasing its technological productivity growth.
　　1- Competitive advantage: By reducing the unit costs of a "given" product by introducing a new production method, the firm would gain a cost advantage against its competitors. That could lead to either a higher profit rate if the price remains unchanged or a price advantage if the cost reduction is reflected by a reduction in price. If completely new products are introduced using a new technology, the firm will expect to enjoy a higher than average profit rate and an access to new markets.
　　2- Monopoly advantage: The owner of a new form of knowledge which produces a **new product or production process** normally applies for a patent, which enables the owner to enjoy a monopolistic advantage until the competitors catch up.   Meanwhile the **expected** profit rate of the technology owner is, probably, above the average market profit rate. In the absence of such expectations, the producer may not have any incentive to finance the risky R&D process in order to introduce new products.

162

3- Defensive strategy: Let us suppose that the competitor of a firm acquires a cost advantage or aims to gain a larger share of the market by introducing new technology. If the firm in question does not take the necessary counter measures, it would risk losing everything. In order to survive the competition, it has to catch up with the other firm either by inventing a competitive technology or obtaining it by transferring it through a patent or a license agreement. In the present economic climate the price of ignorance is annihilation.

What should the characteristic features of a contemporary firm be to maintain a competitive edge in the global market in the long term? In light of the above statements, one can draw the following conclusions:

1- To have access to a labor force endowed with the required qualifications in order to **create** and/or to **efficiently use** the new technology, as well as an appropriate technological and institutional infrastructure.

2- If the required technology is transferred and not created by internal resources, then the firm should have the appropriate capabilities to **adapt** and **further develop** the new technology.

3- Employees of the firm should have access to continuous intra-firm professional training in order to keep pace with the most recent technological developments. Decision-makers in the firm should be open to diverse opinions and promote participation in the decision-making process.

4- Optimum technical or economic efficiency should be one of the major priorities for the firm.

5- The firm's short and long term expectations and goals should be both, rational and realistic in regard to the global facts and developments.

6- The firm must have an appropriate dynamic to take the necessary risks and steps at right time and in the right place.

Technological innovations and consumers

From the point of view of the consumers, generally technological productivity growth provides some major advantages:

1- Cost-reducing technological innovations of "given" products, may reduce the sale price of products, which would lead to increased consumption, cet. par.

2- "Entirely new" or "improved quality" products are introduced in order to provide a greater service for the consumers.

In both cases, the consumers benefit from any technological productivity growth.

Employment and technological productivity growth

Technological innovations usually display two different impacts on employment:

    1- Causing job-losses, thus reducing employment.
    2- Creating new jobs, thus increasing employment.

One frequently comes across cases where people view any new technology with suspicion, and may even display a hostile resistance to change. The underlying reason for this behavior is the fear of job losses, in other words, a loss of income. However, one can observe that while, on one hand, job losses may occur on the other hand the new technology can provide new employment opportunities in many other areas. Which is more beneficial; keeping the "old" jobs with "given" technologies or promoting new technologies and increasing new employment opportunities? Historical developments have already answered this question: the "new" wins over the "old".

Reflections on Technological Productivity Growth

In the subsequent sections, the intention is to demonstrate by using simple models how technological productivity growth affects price, added value, the wage-profit ratio and functional income distribution.

A "Given" Product & a "New" Production Process

All initial production figures used are selected randomly. Since the sole purpose is to analyze the consequences of a technological innovation, this selection of random figures is unimportant.

    Initial values:

$w_t = 100$ TL

$L_t = 500$ workers

$LWC_t = w_t * L_t = 100*500$	$= 50,000$ TL
$OC_t = FC_t + VC_t = 40,000 + 40,000$	$= 80,000$ TL
$TC_t = LWC_t + OC_t$	$= 130,000$ TL

TC denotes total costs, **OC** the total cost excluding wages, **FC** fixed costs (rent, capital goods), **VC** the variable costs excluding wages and such items such as raw-materials, energy, etc., and **(t)** the time. On the income side, **(TR)** denotes the total revenue accrued.

$p_t = 15$ TL

$\mathbf{q_t}$ = 10,000 pieces
$\mathbf{TR_t} = \mathbf{p_t}$ (15)* $\mathbf{q_t}$ (10,000) = 150,000 TL

p denotes price and **q** the quantity produced. The hypothetical size of the profit
($\boldsymbol{\pi}$), the rate of profit (**r**) and the added value (**VA**) will emerge as follows:

π_t = TR$_t$ - TC$_t$ = 20,000 TL
$\mathbf{VA_t} = \boldsymbol{\pi_t} + \mathbf{LWC_t} = 20,000+50,000 = 70,000$ TL
$\mathbf{r_t} = \boldsymbol{\pi_t} / \mathbf{TC_t} = 20,000 / 130,000 = \sim \% 15$
$\pi_t / VA_t \quad = \sim \% 28$ **(share of profit in VA)**
$LWC_t / VA_t \quad = \sim \% 71$ **(share of wages in VA)**

Case-1 Labor(er) saving technological innovation

Let us assume that after the introduction of a new technology the firm continues
to use the same quantity of inputs as before. But the number of employees is reduced from 500 to 300, cet. par. The new technology will alter the rate of profit
and the share of income in terms of VA in regard to wage and profit, assuming
the quantity supplied, price, wage-rate and non-wage input cost remain the
same.

$LWC_{t+1} = w_{t+1} * L_{t+1} = 100*300 = 30,000$ TL
$OC_{t+1} = FC_{t+1} + VC_{t+1}$ = 80,000 TL
$TC_{t+1} = LWC_{t+1} + OC_{t+1}$ = 110,000 TL
$\pi_{t+1} = TR_{t+1} - TC_{t+1}$ = 40,000 TL
$VA_{t+1} = \pi_{t+1} + LWC_{t+1}$ = 40,000+30,000 = 70,000 TL

And;

$r_{t+1} = \pi_{t+1} / TC_{t+1} = \sim \% 36$
$\pi_{t+1} / VA_{t+1} \quad = \quad \sim \% 57$ **(share of profit in VA)**
$LWC_{t+1} / VA_{t+1} \quad = \quad \sim \% 43$ **(share of wages in VA)**

Although the total VA produced is unchanged (150,000 TL), the share of
profit in VA (r /VA) increased from 28 percent to 57 percent. And, although the
real wage level remained the same as before, the share of wages in VA dropped
as a consequence of a labor-saving technological innovation.

Another interesting development is the decline in the VA produced. The reason for this outcome is the reduction in the number of employees, which caused
a decline in the total VA produced (LWC+π).

By assumption, the price was given and remained unchanged. As a result of
the new cost reducing technological innovation, the firm is face with three options:

1- Keeping the price intact and increasing total profits.
2- Increase its competitive strength by reducing price, while keeping the
   profit rate constant.

3- A combination of options 1 and 2.

If the firm prefers to reduce the end price of product, it will gain competitive advantage but at the cost of loosing some profits.

Case-2 An input-saving technological innovation

Let us assume that, after the introduction of a new technology, the same quantity of output is produced with a less amount of inputs but the same amount of laborers, say $VC_{t+1} = 20,000$. Some variables will change, while others remain the same. Those remaining the same are:

$w_t = 100$ TL

$L_t = 500$ laborers

$p = 15$ TL

$q = 10,000$ pieces

$LWC_t = w_t * L_t = 100*500 = 50,000$ TL

And the new values of some of the variables are:

$OC_{t+1} = FC_{t+1} + VC_{t+1} = 40,000 + 20,000 = 60,000$ TL

$TC_{t+1} = LWC_{t+1} + OC_{t+1} = 50,000 + 60,000$ $= 110,000$ TL

$\pi_{t+1} = TR_{t+1} - TC_{t+1} = 150,000 - 110,000 = 40,000$ TL

$VA_{t+1} = \pi_{t+1} + LWC_{t+1} = 40,000+50,000$ $= 90,000$ TL

$r_{t+1} = \pi_{t+1} / TC_{t+1} = 40,000 / 110,000 =$ $\sim \% 36$

π_{t+1} / VA_{t+1} $=$ $\sim \% 44$ **(share of profit in VA)**

LWC_{t+1} / VA_{t+1} $=$ $\sim \% 55$ **(share of wages in VA)**

As displayed above, after the introduction of a new input-saving technology the rate of profit increases from 15 percent to 36 percent, the share of the profit in **VA** changes from 28 percent to 44.4 percent, while the share of wages in **VA** drops from 71 percent to 55.5 percent, cet. par.

Since, by assumption, there is no change in the quantity demanded, there is no reason to reduce the price. However, the firm might wish to use the cost advantage against the competition in order to expand its markets. In this case, the price will have to fall. Lets assume that initially the rate of profit remains the same ($r = \% 15$) and price is subject to fluctuation. Reducing the price from 15 TL to 12.7 TL would not cause any inconvenience to the firm. But, at any price between 15 TL and 12.7 TL would lead to a loss of potential profit, cet. par. Any amount of fall in the price below 15 TL would also imply a decline in the potential share of profits in terms of **VA**.

P_{t+1} $= 12.7$ TL

$TR_{t+1} = p_{t+1} (12.7)* q_{t+1} (10,000)$ $= 127,000$ TL

$\pi_{t+1} = TR_{t+1} - TC_{t+1} = 127,000 - 110,000$ $= 17,000$ TL

$VA_{t+1} = \pi_{t+1} + LWC_{t+1} = 17,000 / 50,000$ $= 67,000$ TL

$r_{t+1} = \pi_{t+1} / TC_{t+1} = 17,000 / 110,000 = \sim \% \ 15$

$\pi_{t+1} / VA_{t+1} = 17,000 / 67,000 = \sim \% \ 25$ **(share of profit in VA)**

$LWC_{t+1}/VA_{t+1} = 50,000/67,000 = \sim \% \ 75$ **(share of wages in VA)**

Case-3   An output increasing technological innovation

Now assume that all the inputs of production, including the employees, remain the same while output increases by 20 percent from 10,000 to 12,000 due to the introduction of a new production process. Once again, some variables will change while others remain unchanged.

First the constants:

$w_t = 100$ TL

$L_t = 500$ employees

$LWC_t = w_t * L_t = 100*500 = 50,000$ TL

$p = 15$ TL

$OC_t = FC_t + VC_t = 40,000 + 40,000 = 80,000$ TL

$TC_t = LWC_t + OC_t = 130,000$ TL

And the changed new values:

$q_{t+1} = 12,000$ pcs

$TR_{t+1} = p_{t+1} (15)* q_{t+1} (12,000) = 180,000$ TL

$\pi_{t+1} = TR_{t+1} - TC_{t+1} = 180,000 - 130,000 = 50,000$ TL

$VA_{t+1} = \pi_{t+1} + LWC_{t+1} = 50,000+50,000 \quad = 100,000$ TL

$r_{t+1} = \pi_{t+1} / TC_{t+1} = 50,000 / 130,000 = \sim \% \ 38$

$\pi_{t+1} / VA_{t+1} \quad = \quad \% \ 50 \qquad$ **(share of profit in VA)**

$LWC_{t+1} / VA_{t+1} = \quad \% \ 50 \qquad$ **(share of wages in VA)**

The new technology causes the initial profit rate of 15 percent to rise to about 38 percent, the share of profit in terms of **VA** rises from 25 to 50 percent, while the share of the real wage in terms of **VA** drops from 75 percent to 50 per-cent, though it remained unchanged. Note that though the real wage does not change, its share in terms of **VA** drops, implying an improved distribution of the functional income in favor of the profits due to the new technology.

Assume that the firm decides to reduce the price of its product from 15 TL to 13 TL to gain a competitive price advantage. This move would cause the profits to fall below the potential amount, but make the firm more competitive.

$P_{t+1} = 13$ TL

$TR_{t+1} = p_{t+1} (13) * q_{t+1} (12,000) \qquad = 156,000$ TL

$\pi_{t+1} = TR_{t+1} - TC_{t+1} = 156,000 - 130,000 \quad = 26,000$ TL

$VA_{t+1} = \pi_{t+1} + LWC_{t+1} = 26,000+50,000 \quad = 76,000$ TL

$r_{t+1} = \pi_{t+1} / TC_{t+1} = 26,000 / 130,000 = \% \ 20$

$\pi_{t+1} / VA_{t+1} = 26,000 / 76,000 \quad = \sim \% 34 \quad$ **(share of profit in VA)**
$LWC_{t+1}/VA_{t+1} = 50,000/76,000 \quad = \sim \% 65 \quad$ **(share of wages in VA)**

In the case presented above, in spite of the price fall to 13 TL, the rate of profit is higher than in the "initial" case. In addition, the firm is in a better position in terms of the share of income. The new technology has not only strengthened the competitive position of the firm, but also raised both the profit rate and the share of profits in terms of total revenue, when compared to the "initial" case.

Case-4   A labor and input-saving technological innovation

In actual economies, new technologies do not "only" save labor or inputs, as discussed in the previous samples, but occasionally even the output as well, "given" the product. Such developments cause the profit rate to rise, which in the long term is the main goal of every firm. As a result, though the real wage rate does not change as in seen in the previous examples, the share of profits in terms of **VA** increases and the functional income is re-distributed in favor of the profit maker, cet. par.

Assume that the initial values are valid and a new technology allows the firm to produce the same amount of output with less inputs and employees.

$w_{t+1} = 100$ TL
$L_{t+1} = 400$ employees
$p_{t+1} = 15$ TL per piece
$q_{t+1} = 10,000$ units
$LWC_{t+1} = w_t * L_t = 40,000$ TL
$FC_{t+1} = 30,000$ TL
$VC_{t+1} = 30,000$ TL
$OC_{t+1} = FC_{t+1} + VC_{t+1} = 60,000$ TL
$TR_{t+1} = p_t * q_t = 15 * 10000 = 150,000$ TL
$TC_{t+1} = OC_{t+1} + LWC_{t+1} = 60,000 + 40,000 = 100,000$ TL
$\pi_{t+1} = TR_{t+1} - TC_{t+1} = 150,000 - 100,000 = $ **50,000 TL**
$r_{t+1} = \pi_{t+1} / TC_{t+1} = 50,000 / 100,000 = \quad \% 50$
$VA_{t+1} = LWC_{t+1} + \pi_{t+1} = 40,000 + 50,000 = $ **90,000 TL**
$VA_{t+1} / L_{t+1} = 90,000 / 400 = \quad 225$ TL / per employee
$\pi_{t+1} / VA_{t+1} = 50,000 / 90,000 \quad = \sim $ **% 55 (share of profit in VA)**
$LWC_{t+1} / VA_{t+1} = 40,000 / 90,000 = \sim $ **% 44 (share of wages in VA)**

Consequently, as the cost of inputs and wages are reduced due to the new technology, the total profits (π_{t+1}) and the profit rate (r_{t+1}) increase along with the share of the profits in terms of **VA**. Functional income is again re-distributed in favor of the profit maker while the real wage rate remains the same.

168

A New Product & a New Production Process

So far, we have analyzed cases where a "given product" was produced by a "new production process". Yet, we know that productivity growth due to the application of a new technology process for the "given products" has only a limited impact on economic growth. For, no matter how much the production costs and/or the price declines due to a new technological process, the markets sooner or later are bound to saturate, which would bring the growth process to an end.

The actual cause of long term productivity and economic growth is technological innovation which introduces "new products", which are, in general, accompanied by "new production processes". The continual supply of new products is the cause of ever rising living standards. Due to technological innovations, both the quality and the quantity of the products placed at the service of the end-users increases along with individual and total wealth. In the absence of such innovation the marginal utility of the "given products" would decline over time which would lead to a decline in profit rates. Eventually economies would in future reach the well-known neoclassical "equilibrium" point and further growth would be contingent upon population growth. New investment would cease and living standards would remain unchanged. In reality this will not be the case due to the introduction of **new products and production processes**, which are the product of **creative mental labor**.

Let us take a closer look at how "new products and new production processes" affect the productivity change in terms of the "aggregate added value" (**AVA**). Assume that in an economy five different products ($q_1, q_2, q_3, q_4,$ and q_5) are being produced. **Q** denotes the quantity of the total output, **Y** the value of the total output and **p** the price of the products.

$q_1 = 3,000, q_2 = 14,0000, q_3 = 5,000, q_4 = 7,000, q_5 = 6,000$ pieces
p = 10 TL
$Q_t = \sum q_i =$ **35,000 pieces** $i = 1,\ldots,5$
$Y_t = Q_1 * p = 35,000 * 10 =$ **350,000 TL**

Further assume that two "new products" are introduced following a technological innovation, say a digital TV, q_6 and a solar energy powered car, q_7. The total supply of the new products is 5,000 pieces and the average price is assumed to be 10 TL, cet. par. Both, **Q** and **Y** will naturally change.

$q_6 = 2,000$ pieces,   and $q_7 = 3,000$ pieces
$Q_{t+1} = \sum q_i = 35,000 + 5,000$ = **40,000 pieces.** $i = 1,\ldots.7$
$Y_{t+1} = Q_2 * p = 350,000 + 5,000 * 10 =$ **400,000 TL**

Thus, the contribution of the new products to the total wealth in terms of value will be **ΔY = 50,000 TL**. Meanwhile, the total quantity supplied (**Q**) increases from 35,000 pieces to 40,000 and the total value of the output (**Y**) increases from 350,000 TL to 400,000 TL.

The related and important questions are: How to determine the prices of the new products? What would be the new profit-rate?

Since the products are **new ones**, there will be no chance to make a price comparison. But, since the owner of the new product will have monopoly rights due to his or her ownership of the patent, the "**expected**" profit rate would, probably, be above the average market rate. This expectation is important for the further development and introduction of new products.

A New Product, a New Production Process and Functional Income Distribution

By assumption, the wage rate of an employee remains unchanged until the next round of wage-negotiations, while the new technologies are being employed. Using simple mathematical symbols the likely impact of the technological innovation in the short term would be as is indicated below, cet. par.:

$$w_{t+1} = w_t$$

But,

$$VA_{t+1} > VA_t$$
$$r_{t+1} > r_t$$
$$\pi_{t+1} > \pi_t$$
$$\pi_{t+1} / VA_{t+1} > \pi_t / VA_t$$
$$w_{t+1} / VA_{t+1} < w_t / VA_t$$

As observed above, the total added value or the total income increases with the employment of the new technology, and the functional income changes in favor of the profits although there has been no decline in the real wage-rate. In other words, every technological productivity growth leads to a re-distribution of functional income in favor of the capital-owner, cet. par.

Introducing a Wage Rise

Deterioration in the functional income distribution for the wage-earners, in spite of a constant real wage-level continues, as a consequence of technological productivity growth, normally, until it is time for the next round of wage-negotiations. The outcome of any such negotiations is uncertain and depends on the bargaining-power of both parties. Let us assume that, after employing the new technology, the new values of certain variables emerge as follows:

$$w_t = 100 \ TL$$
$$L_t = 500$$

$p_t = 15$ TL
$q_t = 10,000$ pieces
$LWC_t = w_t * L_t = 100*500 = 50,000$ TL
$OC_{t+1} = FC_{t+1} + VC_{t+1}$ $= 80,000$ TL
$TC_t = LWC_t + OC_t$ $= 130,000$ TL
$TR_t = 15 * 10,000$ $= 150,000$ TL
$\pi_t = TR_t - TC_t$ $= 20,000$ TL
$r_t = \pi_t / TC_t$ $= \sim \% 15$
$VA_t = \pi_t + LWC_t = 20,000 + 50,000 = $ **70,000** TL
π_t / VA_t $= \sim \% 28.5$ **(share of profit in VA)**
LWC_t / VA_t $= \sim \% 71.4$ **(share of wages in VA)**

And further assume that after negotiations the wage-level increases by 20 percent.

$\Delta w = 20$

The new wage level;

$w_{t+1} = 120$ TL

Naturally, both labor costs and total costs will rise.

$LWC_{t+1} = w_{t+1} * L_{t+1} = 120*500 = 60,000$ TL
$TC_{t+1} = LWC_{t+1} + OC_{t+1}$ $= 140,000$ TL

And,

$\pi_{t+1} = TR_{t+1} - TC_{t+1}$ $= 10,000$ TL
$r_{t+1} = \pi_{t+1} / TC_{t+1}$ $= \sim \% 7$
$VA_{t+1} = \pi_{t+1} + LWC_{t+1} = 10,000 + 60,000 = 70,000$ TL
π_{t+1} / VA_{t+1} $= \sim \% 14$ **(share of profit in VA)**
LWC_{t+1} / VA_{t+1} $= \sim \% 85$ **(share of wages in VA)**

As a result of the wage-rise, though the total added value remains unchanged (70,000 TL), the share of profit in the total value declines from 15 percent to 7 percent, while the share of wages increases from 71.4 percent to about 85 percent. The share of the wage-rate and the profit-rate affect each other in opposite directions.

A New Product, "Monopoly" and the Profit Rate

Assuming that workers at Company-X, using their mental abilities and accumulated knowledge, develops a new medicine against cancer and the enterprise acquires the patent for its exclusive use. Due to its patent ownership, Company-X enjoys a privileged monopoly position against its competitors in the market. In the initial stage, demand for the new product would in all likelihood far exceed the supply. Company-X would set the price as high as the market could bear.

Accordingly, the profit-rate would likely exceed the average the market rate, say by about 50 percent. But in time, the competitors would develop similar products and as the supply increases, the competition will force the price and the profit-rate to decline.

Let's further assume that, given time, the profit-rate falls to the market average-rate, say 10 percent. This hypothetical trend of profit-rate is illustrated on Figure-1, which indicates that at a specific time-period $(t+4)$ the profit-rate declines to average profit-rate for the market. In the absence of competition, Company-X would set the price as high as possible and extract an above average profit-rate, as long as the demand lasts.

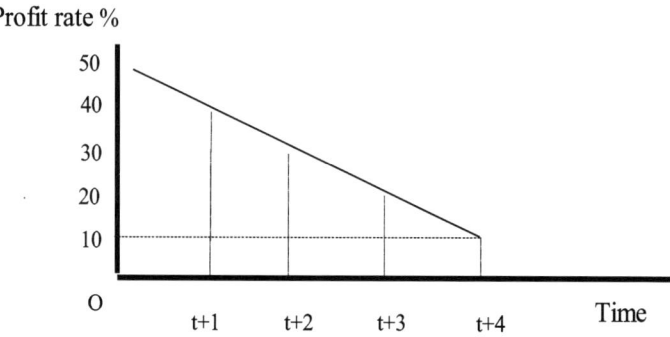

Figure-1   Competition and the changing trend of the profit-rate

Technological Productivity Growth and Price

A "New Product" & a "New Production Process"

Let us assume that the new technology is a new production "process technology" leading to a decline in the production cost of a "given" product. For instance, say that this new process technology reduces the cost of a refrigerator from 100 TL to 90 TL. This situation would give the producer three options in regard to the price:

1- Continue to sell at the same price as before, while increasing the profit-rate:
2- Reduce the end-price to gain a competitive edge on the competition, cet. par.: or,
3- A combination of 1 and 2.

If first option is used, the price will remain unchanged, but the added value created per unit labor-time employed and the profit-rate would increase. If the second option is implemented, both the price and the "potential" profit-rate would decline, while total revenue is likely to increase, cet. par. Third option would provide a combination of these two outcomes.

A producer in a "fair" competitive environment with access to a new process technology leading to a reduction in unit cost would prefer to chose the second option and reduce the end-price. This would be a rational choice paving the way to a larger share of the market and would, in the long term eventually lead to the elimination of the competition, cet. par. The price reduction due to the new process technology would also be in the favor of the end-users, which would lead to a positive "income effect" for them.

A "New" Product, a "New" Process and a "New" Price

Studying the long term trends in economic development, we observe that the real price-level does not show a continuous tendency to decline. One of the main reasons for this is the continual introduction of "new products and processes", which not only causes the profit-rate to rise above the average level but also requires the setting of new prices for new products. The profit-rate of the new product is expected to be higher than average because of the monopoly position and an excessive demand. As long as the monopoly privileges from exclusive patent rights prevail, monopoly profits will continue. But, in time, the competitors are expected to catch up, increase their supply and force the price and profit level to fall.

Studies of long term price indices have to take into consideration that there is a continuous introduction of new products and processes accompanied by "new prices". For example, the latest fad for mobile phones, in practice, provides us with the same kind of service as the traditional house phones. But, nevertheless, they are, technologically, no longer the same, because they contain different features. However, because mobile phones are more expensive than traditional home phones do not mean that the "general phone price" has become more expensive. It would be a great error to apply one price index for all phones, because all phones are not of the same quality. The same conditions apply to the car industry. In 2004 the company's model of car is usually not the same in 2005 or in 2006 in regard to its features and price. Because of these differences in quality, a price comparison would be not only irrational but also erroneous. Due to the above mentioned reasons, there seems to be no price index available at the present time to measure the long term price-level properly.

Other Factors Influencing the Price Level

"Imperfect" Competition

If there was a "perfect" competitive market, as the orthodox doctrines claim, a cost reducing technological innovation of a "given product" would automatically cause the price of product to fall, cet. par. Actually, some prices do tend to decline from time to time as a result of cost-reducing technological innovations and the competitive environment. For instance, in spite of various kinds of qualitative improvement, the prices of computers have, in general, fallen sharply in the last decades. But, there are also some factors which make a price-reduction undesirable for the producers. The main factor is "imperfect" conditions in the competitive markets. If the market is dominated by a few "oligopolistic" companies or by a "monopolistic" seller, cost-reducing technological innovations often do not tend to cause price reductions. The single seller or several sellers may agree on a "no price competition" strategy, which would favor a growing profit-rate per unit output, but is against the interests of the end-users.

Wage Negotiations

Another factor, which might prevent a plausible price-reduction, might be the wage-increase demands of the labor unions. It often happens that the company management meets wage-increase demands positively in order to prevent any possible conflict, which might cause a problem in the supply of their product. The higher the positive expectations of management, the greater the tendency may be to accept the wage-increase demands. Otherwise, a conflict with the unions may cause a serious loss in the markets and profits. Especially in the case of a monopoly or an oligopolistic competition, the management might be inclined to have a more positive approach to a wage rise, because they can easily manipulate the end-price, thus maintain their high profits, cet. par.

In respect to the general price level, a wage-increase in one particular sector may lead to a reduction in the general price level for wage-earners; that is an increase in real income, cet. par. But this development could lead to deterioration in the real income of the wage-earners who, in other sectors, consume the products of wage-increasing sector, cet. par.

The Service Sector

Wage-increase demands by labor unions does not only influence the price level in a related sector where new technologies lead to a cost-reduction, but also the

174

wage and price levels in other sectors, though there may not be any technological innovation, or productivity growth in these sectors. For instance, a wage-rise in an industrial sector due to a technological productivity growth may cause, indirectly, wage-increases in other sectors, such as the private service or the public sector, which due to their internal dynamics display relatively less productivity growth. Public sector services may prove to be an appropriate example. Although productivity growth appears to be insignificant, if not non-existent, this does not stop public servants demanding wage increases. As a result, though the economy may experience a decline in price in certain sectors due to technological productivity growth, the general price level may display a rather different trend, cet. par.

Debt and Interest Payments

Another reason making a decline in the general-price level an unattractive option, is the total debts and the interest payments incurred by the producers. If the price-level falls in a sector due to a technological innovation, this might put the producers, who borrowed to undertake production, in an awkward financial position. Because, the decline in price would result in reduced profits in real terms and increased production costs in regard to the amount borrowed and interests that have to be paid. Therefore, from the point of view of the borrowers a drop in price is not always desirable, cet. par.

The Institutional and the Cultural Infrastructure

Factors such as bad governance of the national economy, inappropriate interventions, and an inadequate institutional and/or cultural infrastructure can also influence the general price level.

The Service Sector and Productivity Growth

So far, special attention has been paid to the use of the term "**product**" to refer to the supply of the producers, which covered both;
1. Physical (tangible) and storable commodities and
2. Non-physical (non-tangible) and storable services.

Some economic textbooks are indifferent to what this distinction really means and when they use these terms in reference to production and exchange they use the terms "commodities and services" interchangeably as though they were the same thing. Yet, if proper attention were to be paid one can clearly see that the traditional economic theories of price formation, production, exchange, growth and trade are, in principle, about "commodities", not "services". Ignoring this distinction in fact demonstrates, unsurprisingly, that service-sector activities, which make up the largest share of the GDP in modern societies, are ignored or overlooked in their economic models.

Actually, for the sake of a proper economic analysis making a separate analysis in the industrial and service sectors in regard to value, price, growth, trade, etc., would be more appropriate and rational. But, unfortunately, this fact seems to have been overlooked since the time of the Classical economists. In fact, service sector activities were often consciously ignored, by amongst others Marx, as being "unproductive" or "parasitic" activities. Only industrial or agricultural economic activities were regarded as "productive". The neoclassical doctrines still appear to follow the same path. In other words, traditional economic doctrines and models, in principle, deal with the production and exchange relations of physical (tangible) storable items that are "commodities", and ignore service sector activities.

Yet, in modern societies, the largest section of employment and output is in the service sector, which was once regarded as "unproductive". The share of the so called blue-collar industrial or agricultural employment is constantly decreasing, though not the per-capita value added by them. As time has passed, the relative portion of the Marxian proletariat has decreased, while that of **"cogniteria"** has increased (Toffler; 1992:p.90).

Is it feasible to measure productivity growth in the service sector as in the manufacturing or agriculture sectors? Can one employ the same criteria in an analysis?

It is hardly feasible, if not impossible, to measure the productivity change in the service sectors. There are simply no physical or storable "things" to measure. Qualitative improvements in the services supplied due to technological progress make the measurement problem even more cumbersome. For instance, how can one properly measure the quality and quantity of the services supplied by a teacher? Or, to give another example, a medical doctor has, nowadays, access to greater variety tools of various qualities than he had 20 years ago. But, how can we compare the "productivity difference" of doctors now and 20 years ago? Should we use a quantitative method? Or would an added value criterion serve the purpose better? Which method could possibly produce more appropriate results in regard to what the students learn?

If one attempted to measure productivity growth or compare productivities anyway, the least cumbersome and most appropriate method seems to be the one based on the added value created per unit-time employed, though it is a far from perfect method.

Conclusion

The main source of long term economic growth and of ever rising living standards is technological progress, which are a product of mental labor. Owing to these technological progresses the long term average profit-rate does not show a tendency to fall below critical limits for continued new investments. Technological innovations not only reduce production costs, but also introduce "new" products and processes, which are crucial for long term growth trends.

The critical sort term impacts of technological productivity growth, e.g., new technologies, are as follows, cet. par.:

1- Real wage rate remains unchanged while **VA** increases.
2- The relative share of real wage in **VA** declines.
3- The relative share of profits in **VA** increases.
4- As a consequence income distribution changes in favor of profits.
5- The income distribution trend is closely associated with the real wage trend.

Functional income distribution changes in favor of profits, because the wage-rate is assumed to be constant in the short term until the next round of wage negotiations. But, in the course of time, as a consequence of positive changes due to technological productivity growth, the demand and pressure for wage increases take a larger piece of the economic pie. The final outcome depends on the bargaining power of two parties and on the specific economic conditions.

"In the long term", the individual and the total wealth and the standards of living tend to increase due to technological productivity growth. In other words, **incessant technological progress appears to be the source of long term and sustained economic growth.**

Developed countries, which are quite well aware of the crucial role played by technological innovation, put great emphasis on education, training and R&D facilities. But, the situation in the other countries which are lagging behind, like Turkey, Peru, Gambia, Vietnam, etc., is rather different. These countries do not necessarily need to make new inventions or innovations to secure growth, at least not for quite a while. Because, if the "available" technologies which are developed and owned by producers in industrialized countries, through patent agreements, can be transferred to the developing countries through appropriate channels, these available technologies could make the same impact as "new" technologies do in the developed parts of the world. Therefore, to benefit from technological innovations, producers in the developing country do not have to pursue the highly costly and risky R&D process as the producers in the developed countries did. The two basic pre-requirements for a successful transfer of technology are first, access to a labor force with the appropriate qualities and second a suitable political-institutional and cultural infrastructure.

But, unfortunately, the global markets for the transfer of technology are far from perfect (Gürak; 2003). Although the adaptive capacity of the recipient country is important, it is not enough for a successful technology transfer. There is an urgent need to formulate new codes of in terms of the transfer of global technology in favor of global competitiveness and, more importantly, in favor of the less developed countries.

Decision-makers in the developing countries should also put great emphasis on increasing the quality of their labor-force in order to make efficient use of the available technologies and to successfully adapt or further develop the new technologies being transferred.

While studying the long term economic growth, technological progress and global economic relationships, it seems important to pay careful attention to and to take the necessary precautions on the following critical aspects:

1- The characteristic features of the global technology markets such as promotion of R&D, ownership, etc.

2- The process of technology transfer and the imperfections in the global technology market (Gürak, 2003). And,

3- The global distribution of the global **VA** produced.

To summarize;

1- Technological productivity growth is the source of long term economic growth.

2- The source of technological productivity growth is **technological innovation**.

3- The source of all technological innovations is **creative mental labor**.

The efficient use of technology requires a **labor-force** with the necessary qualitative endowments. And the long term growth of an economy requires a labor force with **creative mental abilities**.

References

Gürak, H. 1999   On Productivity Growth
                        YK-Economic Review, Dec,Vol.10, No:2, Istanbul.
--- " ---   2000-a   Economic Growth and Productive Knowledge
                        YK-Economic Review;June,Vol.11,No:1, Istanbul.
--- " ---   2000-b   Verimlilik Artışları
                        Verimlilik Dergisi, Eylül-Ekim, MPM, Ankara.
--- " ---   2003   Hidden Costs of Technology Transfer.
                        YK-Economic Review, June, Istanbul.
--- " ---   2004-a   On Value and Price
                        YK-Economic Review, June, Istanbul
--- " ---   2004-b   Büyüme-Teknolojik Yenilik-Nitelikli Emek
                        İlişkisi; Değişim Yayınevi.
Derviş, K. 2011   Küreselleşme, Büyüme ve Gelir Dağılımı
                         www.mfa.gov.tr/data/Kutuphane/yayinlar/
                        EkonomikSorunlarDergisi/sayi27/
                        kuresellesme_buyume_gelir_dagilimi.pdf,
                        2011-09-22
Drucker, P.F. 1995   Gelecek İçin Yönetim. (Managing For Future)
                        İş Bankası Kültür Yayınları No: 327
Vásques, I. 2003-Ed.   Kapitalizm ve Küresel Refah
                        Liberte Yayınları, Ankara.

Others
UNIDO   www.unido.org   Country Industrial Statistics
                            Statistics and Information Networks Branch

6-THE NEOCLASSICAL MARXISTS

Introduction

At the time of Marx, he and many other economists considered labor, in one form or other, given whatever natural resources that were used, to be the origin of all created value. Nothing could be produced without labor's contribution. Thus, every commodity produced could be seen simply as a function of the labor time spent on the natural resources that they used. As William Petty asserted: **"Labor was regarded as the father of (material) wealth, while the earth its mother**." (in Marx,Vol. I,pp.133-134)

With the rise of the Marginalist theory around the 1870s, the approach to the status of labor began to change radically. An ideological struggle against the Marxist's claims provided new premises in economics based on "objective scientific" methodology which led to attempts which claimed to put economics on the same analytical plane as Newtonian mechanics. These new analytical methods suggested that economics ought to be as scientific as astronomy in its precision and its ability to make accurate predictions. Indeed, these attempts succeeded in making economics a "pure" science which became known as "Neoclassical Economics". But, unfortunately, it paid the price of drifting away from the real world and the actual transactions that take place within it. As the Nobel Laureate Ronald Coase once stated, economics became "**blackboard economics**" (Coase; 1991,p.4).

Newtonian mechanics has undergone significant changes since then, but its economic "imitators" remained loyal. Growth and microeconomic theories are still dominated by the neoclassical parable. As in all other parts of the neoclassical tradition, the models are highly abstract and characterized by unrealistic assumptions and mechanical transactions in an artificial world, as if it were a sub-discipline of mathematics rather than an inter-related social science.

In the 1950s, economic analysts re-discovered the crucial and vital role played by technological change, but only as a factor "exogenous" to the growth process. As it was an exogenous factor, it appeared as if it were manna from heaven, its source unknown. An "invisible and mysterious" hand was introducing technological change now and then and then disappearing again without a trace.

Enlightened by these studies and the shortcomings of the economists in the 1950s, some prominent scholars in the 1960s and 70s began to focus their analysis on the diverse qualities of the human resource. It became fashionable to point out the significance and the contribution of a knowledgeable labor force. However the label used was **"Human CAPITAL"** instead of **"Human LABOR"**. This approach which artificially separates "Human Capital" from the rest of the labor force seems to be, amongst others, the major cause of confusion and/or misinterpretation of many of the issues related to the theories of growth or development.

By the way, why is it called Human Capital? Why not **"Intellectual Labor"** or **"Mental Labor"?** Could it be because of ideological clashes? A legacy from the Cold War Era?

Nowadays, a great number of economists acknowledge that both technological change and/or human skill development are irreplaceable (in fact, inseparable) components of the growth process, and should not be left outside the system. The works of Lucas and Romer are clear evidence of this new approach, although it still remains within the framework of the traditional parables. In line with new developments in growth theory which give greater credit to human skills and/or technological change, the inevitable outcome was a modification of the "neoclassical parables" of growth theory. Meanwhile, except for some unnoticed or ignored attempts (Gürak 1991) the "Sanctum Sanctorum" of microeconomic theory was still being kept safe from the infiltration of heretical key concepts such as "Mental Labor" (Human Capital) and "Technological Change".

In the following sub-sections, we shall take a closer look at the works of Lucas and Romer with more emphasis on the latter due to the proper "weight" assigned to **"technological change"** in his analysis of the growth process. According to the conclusions arrived at in previous works, any argument of growth is bound to be sterile unless it is based upon technological change which is "created" by the mental contribution of the laborer (Gürak, 1999, 2000-a, 2000-b). The contributions of Lucas and Romer point out that "human-capital" as the most important factor in their model of **endogenous growth theory**. The substantial similarities of these two well-known scholars to their ideological enemy Marx are striking.

Lucas

In 1988, the Nobel Laureate Lucas, one of the most able minded and prominent scholars of the neoclassical heritage, constructed a **"mechanical"** model emphasizing the formation of human capital as an endogenous factor of growth and attributed a defining role to it. According to Sections 4 and 5 of the model, **human capital accumulation** through formal schooling and learning-by-doing, respectively, was pointed out as an **alternative "engine of growth"** to techno-

182

logical change. Not only growth, Lucas claimed, but also **"relative prices are directed by the human capital endowments"** (Lucas, 1988,p.29).

The Mechanical Model

By "mechanical", Lucas means a model of economic development analyzing an **"artificial world"** populated by **"interactive robots"** as neoclassical economics normally does. The model must be appropriate to put on a computer and run. The relevance of such a model to the real world is, obviously, of minor importance.

Lucas begins his analysis with the assessment of the standard neoclassical growth theory emphasizing "physical capital accumulation" and "technological change". He asks the question whether the typical neoclassical growth model was adequate in order to account for growth and came to a negative conclusion (Lucas,1988,p.6). **"By assigning so great role to 'technology' as a source of growth"**, claims Lucas, **"the theory is obliged to assign correspondingly minor roles to everything else."** (Lucas,1988,p.15).

Then, he considers two adaptations to the standard model to study the effects of human capital accumulation. The purpose is to point to **"an alternative, or at least a complementary, engine of growth to the technological change."** (Lucas,1988,p.17). To do this, he adds "human capital" to the standard model. First, he focuses on a one-sector model in Section 4 of his work with two kinds of capital; physical and human capital. The latter implies a **"general skill level"** acquired through schooling. The only exogenous factor is the rate of population growth.

What can we conclude from his exercises? "Though the model in Section 4 seems capable of accounting for "*average*" rates of growth", says Lucas, "it contains no forces which can account for diversity over countries and over time within a country." (Lucas,1988,p.40). In other words, it says little about the real world.

In Section 5, in his second adaptation of the standard neoclassical growth model, Lucas focuses on **"specialized human capital accumulation"** through learning-by-doing. In other words, all the human capital accumulated is specific to the production of particular goods. Technology and preferences are given, and relative prices are dictated by the abilities of the human capital used. "...**which goods get produced where will also dictate each country's rate of human capital growth.**" (Lucas,1988,p.40). The conclusion is, **".... this account of cross-country differences does not leave room for within-country changes."** (Lucas,1988,p.41).

As mentioned above, **human capital accumulation** to which the capital and the output adjusts endogenously is considered as an alternative engine of growth to technology. The concept of human capital refers to the educated and/or trained part of the wage or salary earning labor force, which has separate features from the **investors' capital**. As the model indicates, Lucas's growth process, e.g., increased value generation occurs in accordance with human capital accumulation. In other words, a more educated or skilled labor force implies a higher rate of growth. Given these distinctive features of the labor force in the growth process, one is then entitled to ask: What is the difference between Marx who claimed that all value is generated by the labor force and Lucas who claims that the engine of growth, thus value generation, is caused by human capital?

Is Lucas a disguised Marxist?

In spite of the key role attributed to "Human Labor", Lucas fails to see any correlation between "human capital" and "technology" (productive knowledge) and states, in his mechanical model, that technology and its level and rate of change is **"something whose determinants are outside the bounds of our current inquiry"**. However, given the features of his model, Lucas earns the right to be called a **"Neoclassical Marxist"**.

Romer

Romer, who in 1990 introduced a new dimension to the neoclassical theory of growth, seems to have a more realistic approach to growth as his model embodies technological change as an endogenous input. At last, technology is acknowledged as an endogenous factor in the growth process. He suggests that something that occurs within the system constantly increases the standard of living. After having asked the accurate question: **"Where is the discussion of innovation, invention, discovery and technical progress?"**, Romer suggests that; **"the most important job for economic policy is to create an institutional environment that supports technological change."** (Romer,1994). Because, new ideas produce new products and new markets which increase the standard of living.

In spite of its ground breaking contribution, Romer's endogenous growth model with technological change at the core of the growth process fails in many respects to reflect actual economic relationships. Inevitably all economic theories are an abstraction from reality. But understanding the nature and extent of this abstraction as well as their unreal reflection of actual transactions is of vital importance in order to properly understand the nature of these relationships in the "real world" transactions and in doing so to be able to design fruitful economic policies. Unfortunately, Romer's model, although a further advanced version of the Neoclassical parable, fails to embrace the important factual ingredients such as the psychological, historical, institutional, cultural and traditional aspects of human relations. But, above all, it **lacks a theory of value and price based on technological change** which is the corner stone of any growth model. In other words, Romer's model is not based on, nor does it offer, **a value/price theory characterized by technological change**. It would be more fruitful if Romer had started with the question: **"Where is the discussion of innovation, invention, discovery and technical progress in the value/price theory?"** In the absence of such a value/price theory, all new theoretical approaches are bound to be sterile.

Nevertheless, we will take a closer look at the model.

The Role of Knowledge

In Romer's own words (1990,p.84), (productive) "**knowledge**" enters into production in two distinct ways in the form of "**new designs**";

1. as new "goods" to produce output; and
2. as addition to the stock of knowledge which increases the productivity of the "Human Capital" in the research sector.

Romer's Inputs and Sectors

Romer's model consists of four inputs and three sectors.

The four inputs are:
1- Capital
2- Labor
3- Human Capital
4- Level of Technology.

The three sectors are:
(i) The Research Sector
(ii) The Intermediate Sector
(iii) The Final Goods Sector

The model studies price-making "**equilibrium**" conditions with monopolistic competition (1990,p.71) whilst ignoring the expected "increasing returns" in real life due to the constant introduction of new designs (technological changes).

There is also the critical assumption of "**free entry**" into, (in reality extremely costly and difficult) R&D activities assuring that "**firms earn zero profit in the present value sense**". (1990,p.73). These two assumptions are totally unrealistic, unnecessary and avoidable but, they are, unfortunately, indispensable ingredients for the neoclassical parables.

In the model, the research sector, which consists only of human capital producing new designs, is the core and engine of his growth model. The higher the human capital stock, the higher will be the rate of growth. Because the new designs, i.e., technological changes are produced by the human capital devoted to the research sector where the stock of knowledge available is the other decisive ingredient. Thus, technological change in the form of the new designs supplied by a **research sector that employs only human capital** utilizing the existing stock of knowledge is "**the true source of economic growth**". Once again, the human capital component of the labor force which produced the existing stock of knowledge and further advances it, is assigned the key role in the growth process, thus placing Romer on the same platform as Lucas; a **"Disguised Marxist"?** or a **"Neoclassical Marxist"?**

Being an able minded scholar, Romer certainly was not unaware that the labor force and human capital were not two separate factors but two sides of the same coin. But, nevertheless, he does not hesitate to make this serious mistake, treating the two as if they were two distinct factors in his analysis.

Why? Is it because he cannot free himself from the shackles of the neoclassical parables?

Let us go back to the four inputs of production and examine them more closely. Inputs one and two, labor and human capital are the two inseparable parts of the same coin, i.e., the labor force, and Romer also admits this. The fourth input, the level of technology, refers to the level of the existing stock of knowledge. This pool of knowledge is produced by the same input i.e., human capital, which constitutes a part of the labor force (Gürak,1999; 2000-a; 2000-b). In other words, knowledge is produced by the educated labor force (human capital) assisted by the eye-hand coordination of a body. Bearing this distinctive feature of the labor force in mind, it would, probably, be more appropriate to study the labor force in two sub-categories;

    1. mental (intellectual) labor, and
    2. manual (physical) labor (see Gürak,1993).

By doing so, we would come to the inevitable conclusion that the **level of technology**, the fourth input, is, in fact, **a product of the labor force, the mental output of the human mind**.

Now we have only two basic inputs of production left, instead of the initial four;

1. the labor force (human resources, qualified and unqualified); and
2. capital goods.

But there is more to it than meets the eye.

In an interview with Joel Kurtsman for the Strategy and Business magazine (1997, 1. Quarter), Romer says that " ... **the law of conservation of matter and energy states we have essentially the same quantity of things we have always had**." This means what we basically do is take the same physical quantity of things, i.e., the raw materials, and rearrange them. In other words, **all kind of goods**, including all the kinds of inputs and outputs that are available, are **rearranged (transformed) raw materials**. In transforming them, we make use of mental and physical abilities of the labor force to rearrange them. Therefore, one can easily claim, like William Petty, Marx and Marshall did a long time ago;

There are only two factors of production; **labor(-er) and nature**.

Is Romer, less of a "disguised" or "neoclassical" Marxist compared to Lucas in regard to his growth model which is based on the knowledge produced by the labor force?

Concluding Remarks

To the surprise of economists, Marx and the neoclassical economists seem to have one very critical aspect in common, namely that;

The source of all value added is the labor force.

The fundamental common shortcoming of both Lucas and Romer is the failure not to base their arguments on a "Value-Price Theory" which includes technological change and human skills. And this failure causes them to build their work on a shaky foundation which allows them to draw inferences more suited to "parables" and fall short in explaining the "real events" that occur in economics.

References

Coase, R.   1991 "The Institutional Structure of Production"
Nobel Lecture, The Royal Swedish Academy of
Sciences. 09-12-1991.

Gürak, H.   1993 An Alternative Price Theory
Unpublished Post-Doctoral Thesis

1999 On Productivity Growth
YK-Economic Review, Dec,Vol.10, No:2, Istanbul.

2000-a Economic Growth and Productive Knowledge
YK-Economic Review, June, Vol.11, No:1, Istanbul.

2000-b Verimlilik Artışları (Productivity Growth)
Verimlilik Dergisi, Eylül-Ekim, MPM, Ankara.

Lucas, R.   1988 On The Mechanics Of Economic Development.
Journal Of Monetary Economics, July, 1988,342

Marx, K.   1976 Capital, Vol. I
Penguin Books.

1977 Capital, Vol. II
Lawrance & Wishart, London.

1981 Capital. Vol. III
Penguin Books.

Romer, P.M. 1990 "Endogenous Technological Change"
Journal of Political Economy, Vol.98, October.

7- FINAL REMARKS

The ambition "**to have more**" possessions whether in the form of money or products, for many individuals appear to be their main goal in life and the source of all their happiness. Individuals as the consumers of products have to learn to say "enough" and attempt to put a rein on this unbridled ambition "to have more", which undermines the humanity inside us. After all, when we die we cannot take our possessions with us.

The emphasis in the six previous sections of this book dealt with the "**creative mental abilities**" of the laborers. That is natural because it is basically these "mental abilities" supported by physical labor which create the new products and production processes demanded by the markets. Accordingly, it is again the mental abilities of men that help to organize the economic system and run it efficiently in accordance with the rules. In addition, all kinds of economic decisions and transactions involve the use of man's mental abilities. If these mental abilities were not as developed and creative as they are, human beings would not have material living conditions that are different from any other creatures. This being the case, human effort would simply be confined to the finding of food, shelter and protection. Not only would there be no technological progress, no scientific advancement, no cultural activity nor any of the other diverse activities created by man's mental ability. The sole purpose of all creative mental activity is to improve living conditions and to make the world a better place in which to live.

The structure of economic relationships and the types of transactions undertaken has constantly been changing and adapting for centuries, even millennia, come to terms with new conditions as they arise. In our age, consumers have access to an unprecedented amount of variety, quality and quantity in their products. But, there seems to be a disturbing defect in the present economic system which favors, primarily, the interests of economically powerful groups; a feature also clearly observed in previous economic systems. These observations clearly indicate that it is not the working people who make up the largest percentage of the population who have the greater political power, but the much smaller group of economically powerful people.

It is also observed that it is normally these economically powerful groups which first reap the benefits of economic growth, while the majority, the working people, consistently suffers first and hardest in the event of an economic decline. Yet, sustaining and maintaining the stability of the economic system depends heavily on the consumption behavior of the majority, i.e., the working people who constitutes the majority. In other words, consumption demand not only has to continue uninterrupted but by pushing all the limits have even to increase. In order to achieve this goal, consciously or unconsciously, working people have to compete fiercely for higher paid jobs. As mainstream economic theory states, consumer behavior has to be "selfish" as well as "economically rational".

There are endless and countless attempts to encourage this "selfish" and "economically rational" consumption most of which employ "brainwashing" techniques and "diversion tactics". Positive humanitarian feelings have been on the decline whilst these "selfish and economically rational" economic decisions are being praised and promoted in order to increase the hunger for more possessions. There are certainly some exceptions but most people seem convinced that the more they consume the happier they will feel. Only a few seem to question and discuss the merits and demerits of this consumption explosion and worry about the potential cultural, social and psychological consequences. In the meantime, the number of the discontented and critical opinions has been gradually but steadily increasing. Some of these critics point to the lack or insufficiency of "moral values" in the system which may lead to the demolishing of social and cultural mores. There are some scientific studies examining increased selfishness and the deterioration in human behavior which give support criticisms such as these. However, there is also an increasing number of people who are showing a more caring, benevolent and respectful attitude towards the world's natural resources.

Since the main purpose of all economic relations and transactions is to improve welfare and to benefit mankind, we have to reconsider the merits and demerits of the economic system we live in. If necessary, we should not hesitate to change our habits and try to live in harmony with other people and nature.

What Kind of Economic Order?

If one of the major characteristics of the economic system in which we live is the "destructive competition" between enterprises as Marx and later Schumpeter pointed out, the other surely is the insatiable appetite of many individuals constantly fuelled by these same competing enterprises that we should have "have more and more". In fact, according to the internal logic of the economic system the consumer and the producer complement each other; the one cannot do without the other and both share many common interests. While the producers do their best to destroy their competitors and wipe them out of the market, the consumers strive to maximize their possessions, expressed as "maximizing utility", according to the prevailing economic jargon. This requires the consumers to compete with each other for higher paid jobs and their incumbent benefits. The higher the income the more possessions, or, to use the economic jargon, the higher the utility obtained. So, there is a constant competition not only amongst producers for a bigger share of the profits but also amongst consumers for the higher paid jobs in order to be able to "have more" possessions.

At present, human beings have access to an unprecedented amount of products in terms of variety, quality and quantity compared to 100 or even 20 years ago. Yet, they don't seem to be satisfied with what is available and demand to "have more" of everything. The word "enough" in terms of the variety and quality of products does not seem to be in the vocabulary. The quantity available is a different matter based on the inequalities inherent in the income distribution system. Even if no "new" products are developed and introduced into markets in the future, those available would be enough to satisfy our needs and desires. But, we still don't know how to say "enough"; or rather, from the point of view of the profit maximizing enterprises, it is undesirable to say "enough".

As Ruben (2011) stated:

> ".. we are not spending our time for ourselves but to work and to earn more (and to have more possessions). Because, the system built on the principal of maximization is enforcing us to do so… The present system produces individuals rich in terms of possessions, but poor in terms of time (and benevolence)". (Ruben, 2011, p.71-72)

Is it possible to transform a society characterized by "selfish and economically rational" behavior into an unselfish, benevolent kind and generous society?

Certainly this kind of transformation is, in principle, possible. There is no limit to what human beings can achieve with their "creative mental abilities", if they really want to. But, in order to be able to realize a successful transformation, society, meaning both the producers (capital owners) and the consumers (individuals), will first have to change their habits and behaviors with respect to profits and possessions; and then to adapt themselves to the principles of unselfish and benevolent thinking and living.

In principle, the transformation of society is possible but; where should the transformation begin? And how is it to be achieved?

From the Lordship of Possessions to the Lordship of Humanity

It is the purpose of all economic relations and transactions to maximize the quality of life and human happiness. In order to be able to improve the quality of life without the obsession with possessions, the habits and the ambitions of human beings have to change first. The ambition to possess more and more in a "selfish and economically rational" manner a là the neoclassical doctrine has to change by giving priority to the improvement in the quality of one's personal life, caring more for the environment as well as behaving kindly and generously towards others. In order to be able to achieve these goals we have to act with greater care and benevolence to others, a form of behavior that should be encouraged and praised by the establishment. In other words, we need some form of formal and informal education which is able to improve our perception of the quality of life paving the way for more unselfish and benevolent behavior.

A conscious person who not only acts rationally but also with benevolence towards others, and is aware of the consequences of his/her decisions can be described as an "ethical" person whose manners are suited to the generally accepted moral values of society. But the term "ethical" is a highly subjective term which varies in accordance with the cultural, the religious, and the normative values of any community. The related and critical question which springs to mind is: What criteria could one set in order to evaluate the "ethical" behavior of people? In the past, religious beliefs played a critical role in determining and evaluating "ethical" economic behavior. For example, it was once considered that being rich and having possessions in excess of those required to have a decent life was a mortal sin. The ambition to possess more was not regarded as being in line with God's will. Nor was extravagant consumption or thriftiness approved by the clergy.

Yet, in so called "modern" communities, everybody is being encouraged through advertising to "have more" because "having more" makes you "happier". In contrast to the practices and beliefs in the past, at present, we are being indoctrinated by profit driven enterprises to accept that consuming or possessing "more" is the only source of happiness in our modern lives. Through the means of commercial advertising, consumers are constantly being brainwashed to believe that having more possessions leads not only to "more" happiness but also "more", i.e. a higher, status in the community. Those who possess less than others often feel downgraded and unhappy as they have a standard of living below the standards of their modern counterparts. The major purpose in life for many people seems to be to possess more and to consume more. As Fromm (2003, p.139) pointed out, it is a widespread belief in modern societies that since man has existed, the desire "to possess" is a human characteristic and therefore cannot be altered... This dogma is accepted as given in contemporary society and determines our upbringing as well as the way we do things. However, this behavior is in fact nothing more than a tendency to adjust to the prevailing social mores.

If human beings are ever to become free and to cease feeding industry by pathological consumption, a radical change in the economic system is necessary: *we must put an end to the present situation where a healthy economy is possible only at the price of unhealthy human beings.* The task is to construct a healthy economy for healthy people... *The first crucial step toward this goal is that production shall be directed for the sake of "sane consumption".* (Fromm, 2012, p. 143)

Fromm is right in his criticism stated in the paragraph above but there is another side to this coin. Can a producer who is not profit driven and unselfish in regard to his economic decisions survive the competition in the long term? Or, in the long term, what are the chances of a producer who does not strive to destroy the competition being successful?

From the Producer's Point of View

Using the monopolistic or oligopolistic methods of organization in production and distribution are undesirable in the competitive markets because of inherent features which lead to an inevitable distortion in the market. However, obtaining the patent rights for a new technology automatically gives the technology owner a monopolistic power over the market for a specified period of time. The next "economically rational" (?) step expected from a monopolistic technology owner is to gain the control of the market and, if possible, drive away, e.g., destroy, all the competitors in the market. In fact, the elimination of the competitors from the market or defending one's present position in the market is the major goals of all enterprises that use new technology. This hypothesis is valid in the production of all physical goods. There is a fierce competition even in service production sector but it displays some different features. Competing enterprises in the service sector can manage to survive in the long term without wiping out the competition provided that they make a profit.

The type of enterprises with "unselfish" competitive behavior will be the ones which stand the highest risk of being eliminated from the market. An enterprise in the competitive market cannot "live and let live"; the one will always attempt to destroy the other. That is the nature of the competitive market system. As long as this economic system prevails, one enterprise cannot avoid or overlook the others' actions. However some measures can be taken to reduce the negative impact of the system. Four different types of measures are stated below, a list which can of course be increased.

1- "Investment Credit to the Employees"

It is well known that, the present so called "free market" system is quite gener-ous in providing various kinds of financial incentives ranging from tax exemp-tions, subsidies to low cost credits, etc. Such incentives help to make the already rich capital-owner richer and the economically powerful groups more powerful, and are often accompanied by a deteriorating income distribution. This system which favors and consolidates the position of the economically powerful groups can to some extent be reversed, by changing this system with one more favora-ble to the employees. This could be achieved by the giving of some financial subsidies and incentives to the employees with the sole purpose of their using it in some required future investment. To put it another way, the economic incen-tives which up to now are protecting and encouraging the interests of the capital-owner can be changed to share some of the accruing benefits with the workers as a kind of "investment credit". Thereby the laborers' share of the dividends can be increased which in turn would expand the workers' ownership in the enter-prise and increase their participation in the decision-making process in regard to the related economic issues. In such a system, in periods of downswing, the cap-ital-owners would not be able to fire the employees easily while the laborers would be more prone to accepts wage-cuts, and enjoy more of the profit in times of expansion. Another interesting feature of such a system would be the im-provement in functional income distribution in favor of the employees.

The benefits of a new financial system can be briefly summarized as fol-lows:
1. Ownership would be expanded.
2. Functional income distribution improves in favor of the laborers.
3. Economic democracy at the production unit increases.
4. The political power of the laborers increase.
5. Laborers show less resistance to wage cuts in times of crises.
6. Laborers would support measures to increase productivity.
7. Waste in regard to inputs would be reduced.
8. Loyalty to the enterprise increases.

2- Encouraging and Promoting Third Party Ownership

In the present economic system 'some' of the wealthy with income in excess of their needs use this surplus in the financial markets to obtain an interest rate; a process which can be described as "making money from money". These in-comes in excess of the consumption requirement constitute a pool of financial funds which are used, through financial intermediaries, to finance credit requests of both the individual and the enterprise. As a result of such credit transactions, the financial intermediaries receive a certain rate of interest, some of which is paid to the money-holders. This system enables the money-holders to increase their money-holdings, i.e., making more money from money without making any direct or concrete contribution to the product supply.

Though the number or proportion of persons with huge sums of money-holdings is relatively small, the size of the income from their "non-productive" money deposits reaches incredible amounts. These "non-productive" money-holdings which are kept as deposits can be utilized in a more economically effective and rational way to directly finance the supply of products by actively becoming involved in production and assuming the risk as well as the responsibility for a particular production process. As a result, not only will the suppliers of the products will have access to investment funds at a lower cost but also property ownership will be expanded leading in turn to a more democratic economy. (For a more detailed study of the "interest rate" see Section two of this book).

3-"Profit sharing"

The individuals who make a direct contribution to the supply of various goods and services can be divided into two major groups: 1- Employees; 2- Employers; or, alternatively, 1- Laborers; 2- Capital owners or capitalists. As we all know, the individuals of the first group, that is the laborers, obtain wages for the labor services they supply while the capitalists receive profit in return for their investment.

As long as demand for the supplied product does not show any tendency to decline, production will be sustained, so laborers will continue to earn their wages and capitalists their profits, cet. par. It is high time to reconsider the merits and demerits of this functional income distribution whether it is a "rational", "fair" and "egalitarian" system and to make changes when necessary. A thorough discussion of the following proposal might produce globally useful results for all economies. The proposition we will assert claims that:

"Since the services of laborers are an indispensable and inseparable input of all production, then they should be entitled to obtain or to acquire 'a part' or 'a portion' of the profits obtained by an enterprise."

If the laborers are allowed to obtain or to acquire 'a part' or 'a portion' of the profit regardless of share ownership, they would be better motivated to increase their productivity as well as to minimize the waste of the time and the resources which would, in turn, improve the competitive strength of the enterprise and increase the rate of profit, cet. par. In addition, the inequality of the distribution of the functional income, the ancient and unresolved headache of the capitalistic economic order, would improve, to some extent, in favor of the laborers. As a result, both the demand for products and rate of employment is expected to increase, cet. par.

4- Independent Units of Technology Supply:

The most costly research processes in the most dynamic sectors of the economy is to invent and develop "new products" and "new production methods". This is usually carried out by huge globally operating enterprises with strong financial facilities. The emergent new technology is normally protected by patent rights which give the patent holder a monopolistic power in the market for a specific period of time which leads to monopoly profits as well as the right to control the quantities that will be supplied.

A new economic order which allows the establishment of "**independent technology producing profit units**" would have a great positive impact on the functioning of the markets which would in turn benefit the consumers by increasing global competition. Instead of a monopoly, there will be a basis for global competition in the markets by manifold enterprises. Because these independent technology producing enterprises which are driven by the profit motive would be anxious to sell their patented new technology to as many customers as possible, i.e., enterprises, in order to maximize their profits. As a consequence there would be no single enterprise with a monopoly power that is able to extract monopoly profits by determining the market price or the quantities supplied, or both. Thus, the consumers will be able to obtain and use the new products at a relatively lower cost.

Another positive contribution of such a system would be the improvement in the global distribution of income in favor of the relatively less developed countries and their enterprises, provided that no new and unforeseen "market distortions" are introduced by powerful global enterprises.

Another important and related issue is the prevention of a "brain-drain" from the less developed to the developed countries and their enterprises. In the case of the lack of a sufficiently qualified labor force, global competition with an advanced country's enterprises would only be illusory even in the case of the independent technology producing enterprises.

196

Competition vs. Coexistence

An Old (Ancient) Turkish Professional Organization: The "Ahi" Institution

Could an Old Turkish Professional Organization be a Model for present day economic transactions?

The so called "Non-Government Organizations" (NGOs) often play a critical role during the process of the development of a nation or a country. The Turks had a similar organization known as the "Ahilik". An organization of "small scale enterprises" similar to the western "Guilds" which played an important role for hundreds of years in the organization of production and the development of Turkish society. The roots of the "Ahi" organization dates back to the period of time when Turks were living in Central Asia. Both of the Turkish Empires, the Seljuk's and the Ottomans had benefited to a significant degree from the "Ahi" organization not only to preserve social and economic stability but also to foster the development process. The organization and its influence were encountered not only in the towns but also in the countryside. After the conquest of Istanbul in 1453, the Turkish Emperor, Sultan Mehmed II imposed institutional changes which resulted in the decreased function and importance of the "Ahi" organization and it gradually turned into an ordinary "professional organization".

What was unique about the Ahi organization? Which features made it different from other Guilds? Which were its aims? And could the Ahi organization achieve their goals?

The primary purpose of the Ahi organization was to build a "perfect society" living in prosperity and security, and consisting of individuals with high "moral values". Not only in regard to the conduct of the professional members of the organization but also in regard to the products that they supplied. Both had to be "perfect". In order to reach this goal, the members of the Ahi organization were being trained and educated to be individuals with "high moral values and virtues". The training and/or education they acquired can be headed under three groups:

1. Moral training.
2. Military training.
3. Professional training.

From the list above we can see that the members of the Ahi organization did not only receive professional and educational training. Moral education and training in accordance with the prevailing religious values were, at least, as important as the professional education and training. Military education and training was necessary in order to protect the society in which they lived in from external aggression as well as to maintain social order. After the consolidation and the emergence of the Empire, the military functions expected from the Ahi organization were changed and more emphasis was placed on the professional and moral aspects of their education.

According to its basic principles, the three restrictions or constraints of an Ahi, a member of the Ahi organization, are:

1. His eyes should not see the things forbidden by his religion.
2. His mouth should not be used to commit a sin.
3. His hands should not be used to oppress others (cruelty).

The three benevolent functions of an Ahi that are "unconstrained (unrestricted) are:

1. A door that was open for guests.
2. A purse for brothers (other Ahi members) in need.
3. The table (food) for the hungry.

(www.ahilik.gen.tr/kavram/misyon.html, 2012-01-26)

The principles of "professional ethics and manners" of an Ahi institution are:

- To be good tempered and decent.
- Not to covet the things of others.
- To be generous, kind and beneficent.
- To be affectionate towards the young; to behave well and be respectful towards the old.
- To be modest; to avoid being arrogant or proud.
- To be benevolent and to wish an increase in prosperity and goodness for all.
- To carry out one's work sincerely, with generosity, kindness and a smile.
- Always search for faults and errors in yourself before accusing others.
- To protect and look after his subordinates and servants.
- To avoid wicked talk and manners.
- To behave genuinely; to be trustworthy.
- To be friends with wise men; to consult with friends.
- To conform to customs and traditions.
- To be satisfied with "little"; to be grateful with plenty and to share it with others. (www.ahilik.gen.tr/gorgu/temel.html,2012-01-26)

The Lessons to be Drawn from the Ahi Institution

Attempts to reincarnate an institutional or an organizational setting which proved "successful" a 1,000 years ago and expect "the same" success in our age would be, to say the least, meaningless. Since that time not only has Turkish society undergone drastic economic, cultural, institutional and political changes but also have many other countries. The requirements of the technological and labor force are different, as are those for the individual and society. Therefore, the answer to the question whether the old style Ahi institutional setting can produce the same results is obviously, no it can't. However, one can always draw some lessons from past experiences and can make use of them by introducing them after some appropriate amendments adjusted to the present conditions. Some or all of the above principles can be relevant at this moment in time. For example, those who subscribe to such principles may be rewarded by encouraging such benevolent behavior. An important detail that should not be overlooked is that the principles of the Ahi organization were intended for small scale production enterprises.

From the Consumers Point of View

Compared to the individuals in the supply sector, consumer expectations and interests are somewhat different which might make it relatively easier to find a remedy to the problem of "selfishness." Consumers as individuals tend to reveal "unselfish, contented and benevolent" behavior in contrast to the expected neoclassical doctrines' "selfish" and "economically rational" behavior. This kind of "anti" neoclassical" economic behavior may even contribute to the greater personal happiness of the consumer, assuming that their "basic needs" are met in regard to the development level of the society in which they live. In any case, benevolent and conscious individuals always appear to avoid extravagant consumption, the pollution of the environment and arrogant attitudes.

In our society there are always some people around us who display benevolent behavior, and their number seems to be on the rise. Moreover, there is also a cultural and philosophical infrastructure with "various kinds of moral values" originating from both, the East and the West. If "benevolent" consumers rearrange their consumption priorities, the infrastructure of the supply conditions and producer priorities will have to change. As a result, as Fromm (2003, p. 212) once pointed out, the production relations will have to change and become more favorable to consumers with "rational" choices, instead of favoring producers who attempt to maximize the quantities they supply and the profits they realize.

The Quality of Life and Things

A group of graduates pay a visit to one of their Professors at the university.

The conversation soon turns into complaints about the stress at work and in their lives. The Professor who wishes to offer coffee to his guests goes to kitchen and after a while returns with a big thermos full of coffee and different kinds of cups made of china, plastic, glass and crystal some of which look cheap, some expensive and some exclusive. After each guest has picked up a cup, the Professor says:

I don't if you've noticed but, all the expensive looking cups have been picked up, leaving behind only the cheap and simple looking cups. Though it is quite natural to desire the best for yourself, this choice in fact is the source of all your problems. Believe me; the cup itself adds nothing to the quality of the coffee. Usually, a more expensive cup sometimes disguises what we drink. Actually all we wanted was to have a cup of coffee, not the cup; but consciously you have chosen the better cups and then started looking at each other's cup.

Think about this:

Life is a cup of coffee. Our job, our money and our position in society are simply the cups. They are just the means "to have a good life"; and the cup we pick up neither determines the quality of life nor helps us change it.

Sometimes we forget to enjoy the coffee and focus only on the cup.

Enjoy your coffee.

The happiest people are not the ones who possess the best things.

Source: Internet

Concluding Remarks

Thousands of years ago, men first met their immediate personal and household requirements by producing various kinds of products with "use-values" or "utilities", by utilizing their mental and physical abilities. They then began to exchange their excess products with each other thus increasing the total supply of products, e.g., the total added value and the quantities consumed. During the past hundreds of years mankind continuously improved his personal living standards as well as the total material living standards of the community by constantly improving his personal mental abilities and making use of the resources available.

The main source of long term economic development with its increasing variety, quantity and quality of products has always been a result of human mental ability. From time immemorial mankind has climbed high up the economic ladder in line with the ongoing cultural, technological, scientific and organizational developments that have occurred in the ensuing period. In other words, the present level of material well-being is due to the constantly introduced technological innovations which are the products of "creative mental labor". This level of material well-being was the result of the use of technology by laborers who in their turn made use of the available and appropriate educational and institutional developments to improve or develop new technology. As we clearly observe, human mental ability has been efficacious in every stage of development effective since the beginning of human existence. And the principle purpose of development has always been the well-being of mankind, i.e., improved material living conditions. However, the latest stage of economic development the so called "free market economy" or rather "the capitalist system" has been accompanied by many unforeseen economic and non-economic imperfections.

The nature of the present economic system and the expected economic transactions taking place within the system resulted in the emergence of a society where individuals, in general, do not seem to pay sufficient attention or care either for the well-being of "other" human beings or for nature. There are plenty of "self-centered" and "economically rational" individuals a lá the neoclassical doctrine who seem to go through life like mechanical robots caring only about their own well-being, having "faith" in man-made "objects", i.e., the new "holy value". Human beings are constantly being indoctrinated by profit driven commercial enterprises and their efforts appear to be gaining ground. The system successfully and increasingly creates men addicted to obtaining "objects" of varying kinds and quantities. Instead of man being "the lord of objects and money", the opposite seems to be taking place; "the man-made object and money" seems to be the "new lord of man".

The Fable of the Ant and the Grasshopper

In this well-known fable the industrious and cautious Ant works hard during the summer and stores the food that he collects, while the grasshopper plays music and sings songs. When the winter comes the Ant has plenty of food but the Grasshopper who finds no food to eat knocks on the door of his neighbor the Ant and asks him if he can borrow some food and he will pay him back the next summer.

The Ant asks: "What did you do during the summer?"

The Grasshopper replies honestly: "I played music and sang."

The Ant replies:

"Since you played music and sang during the summer, it's time to dance now."

The main lesson we are expected to draw from this tale is "to be as industrious and as cautious as the Ant and to save for harder times." Those who are lazy and only lead a life of pleasure as the Grasshopper did will sooner or later end up in dire straits."

Yet, we can interpret this tale in a totally different way. When we read this tale to a child without making any comment, the child probably would regard the Ant as merciless and unfeeling and be angry at him for not helping the Grasshopper. The child probably would go further and admire the Grasshopper for being entertaining, playing music and singing songs and perhaps would even try to help the Grasshopper himself.

Although this tale may have been read and told at home, in kindergartens and in the primary schools thousands of times the adults (parents and the teachers) never mention how selfish and merciless the Ant is and how the Grasshopper loves "the joy of living".

Let's reconsider the tale; the Ant apparently represents the selfish individual who is controlled by the capitalist system. He is ambitious and anxious to earn more and he works hard in order to reach his goals, but he lacks such values as empathy, compassion and friendship. The Grasshopper who is cheerful, joyous and enjoys his life but who also is lazy and improvident. He is definitely not an individual of whom the capitalist system would approve. Yet, a healthy individual is one who works hard but who spares time for himself or herself to enjoy life.

Let's assume that at the end of this well-known tale, when the Grasshopper visits the Ant to ask for help, the Ant invites him in, they eat dinner together and then the Grasshopper plays his music and sings his songs, greatly entertaining the Ant. Then during a friendly chat the Grasshopper admits that in future he should be more cautious, and in the meantime, the Ant says that playing music and singing songs makes a person feel better, and so they become good friends.

Don't you agree that this would be a much better message for children?

E.B. Ruben, 2011, Iktisadın Unuttuğu İnsan, p. 80-81

In order to be able to make radical "humanitarian" changes in order to reverse the present economic relationships and behavior, mankind has to undergo some drastic changes in regard to its thinking and actions. The "selfish behavior" of mankind will have to be replaced by "unselfish behavior" and the "economically rational transactions" a lá the neoclassical doctrine replaced by "more benevolent actions, more affection, more tolerance and more respect for social values". The concept of "richness" will have to change and give way to "having more friends and being able to share more personal objects with others" instead of constantly accumulating personal wealth in terms of the objects possessed and money.

Bibliography

Fromm, E.   2003 Sahip Olmak ya da Olmak
                       Arıtan Yayınevi, İstanbul. Çev. Aydın Arıtan
From, E.   2012 http://books.google.com.tr/books?id=JvG85s966ko
                       C&pg=PA137&hl=tr&source=gbs_toc_r&cad=
                       4#v =onepage&q&f=false, 2012-05-19
Gürak, H.   2011 İktisat,
                       Genesis Kitap, Ankara.
Orman, S.   2010 İktisat, Tarih ve Toplum
                       Küre Yayınları, İstanbul.
Ruben, E.B.   2011 İktisadın Unuttuğu İnsan
                       Bağlam Yayıncılık, İstanbul.

Web Sites:

    www.ahilik.gen.tr/kavram/misyon.html, 2012-01-26
    www.ahilik.gen.tr/gorgu/temel.html, 2012-01-26